T0290851

ADVANCES IN PLASMA
PHYSICS RESEARCH

VOLUME 7

ADVANCES IN PLASMA PHYSICS RESEARCH

Additional books in this series can be found on Nova's website
under the Series tab.

Additional E-books in this series can be found on Nova's website
under the E-books tab.

ADVANCES IN PLASMA PHYSICS RESEARCH

VOLUME 7

FRANCOIS GERARD

EDITOR

Nova Science Pubishlers, Inc.
New York

NOTICE TO THE READER

The Publisher has taken reasonable care in the preparation of this book, but makes no expressed or implied warranty of any kind and assumes no responsibility for any errors or omissions. No liability is assumed for incidental or consequential damages in connection with or arising out of information contained in this book. The Publisher shall not be liable for any special, consequential, or exemplary damages resulting, in whole or in part, from the readers' use of, or reliance upon, this material. Any parts of this book based on government reports are so indicated and copyright is claimed for those parts to the extent applicable to compilations of such works.

Independent verification should be sought for any data, advice or recommendations contained in this book. In addition, no responsibility is assumed by the publisher for any injury and/or damage to persons or property arising from any methods, products, instructions, ideas or otherwise contained in this publication.

This publication is designed to provide accurate and authoritative information with regard to the subject matter covered herein. It is sold with the clear understanding that the Publisher is not engaged in rendering legal or any other professional services. If legal or any other expert assistance is required, the services of a competent person should be sought. FROM A DECLARATION OF PARTICIPANTS JOINTLY ADOPTED BY A COMMITTEE OF THE AMERICAN BAR ASSOCIATION AND A COMMITTEE OF PUBLISHERS.

Additional color graphics may be available in the e-book version of this book.

Library of Congress Cataloging-in-Publication Data

ISSN: 1544-8231

ISBN: 978-1-61122-983-7

Published by Nova Science Publishers, Inc. ✝ New York

CONTENTS

PREFACE

This new book presents and discusses current research in the field of plasma physics. Topics discussed include the dynamics of the plasmasphere; dephasing in Rydberg plasmas; plasma turbulent transport modelling; nonlinear dynamics of self-guiding electromagnetic beams; supersonic molecular beam injection in fusion plasma and an experimental study on tandem mirror edge plasmas.

Chapter 1 - The plasmasphere is the high altitude extension of the ionosphere. To ensure accuracy and reliability of communications, navigation and other satellites stationed in this region, there is a considerable interest to understand the plasmaspheric environment and its dependence on external parameters. Dynamical simulations have been developed at the Belgian Institute for Space Aeronomy to simulate the deformation of the plasmasphere during geomagnetic substorms and other variations in the level of geomagnetic activity. These simulations are based on the mechanism of plasma instability and on the E5D empirical electric field model. The results of these simulations are compared with satellite observations of the plasmasphere. Comparisons with observations of the IMAGE spacecraft launched in March 2000 are especially useful to understand the dynamics of the plasmasphere since the satellite provides the first global comprehensive images of the Earth's plasmasphere. The EUV (Extreme UltraViolet) instrument gives intensity maps of the 30,4 nm emissions of Helium ions integrated along the line of sight. It revealed the global behavior of the plasmasphere and its dynamics influenced by the solar wind and interplanetary magnetic field variations. New structures in the plasmasphere such as shoulders, channels, notches and the formation of plumes were discovered. The four Cluster spacecraft launched in 2000 also provide useful data to compare with the simulations.

Chapter 2 - In Coulomb systems, bound states can be formed which are obtained from the solution of the two-particle SchrÄodinger equation. Due to the interaction with the surroundings, transitions between these stationary states are possible so that a finite lifetime for the bound states is observed. In particular, Coulomb interaction as collisions with other charged particles as well as the interaction with the transverse part of the Maxwell field, leading to the emission and absorption of photons, can be considered using the method of thermodynamic Green functions. In a more general context, the finite lifetime of the bound states can be treated as the destruction of phase coherence due to the influence of the surrounding bath. On the other hand, decoherence and dephasing are concepts to characterize the damping of off-diagonal elements in the density operator. At present, decoherence is an important concept to investigate open quantum systems, which are in contact with a bath so that irreversible time evolution is possible. Authors will consider Rydberg atoms which are weakly bound electrons in highly excited atomic states. Measurements of the transition rates for Rydberg states with principal quantum number n = 10 - 40 due to Coulomb interactions are reported. They may be considered as mesoscopic systems. An interesting question is the transition from atomic quantum states to a quasiclassical wavepacket of electrons in bound states. Recently, experiments have been performed to investigate this basic question in atomic physics. A more applied aspect is the possibility to construct a quantum computer basing on the coherent superposition of quantum states, for which also the dephasing time is an important limiting quantity. Authors interest is also to find the connection to the standard description of mesoscopic system, such as semiconductor structures, which has been elaborated a few years ago, see, e.g. Imry. In this field of research, some new and interesting experimental results are obtained recently. In particular, they will check the applicability of the Imry's approach for the dephasing time given there on the special case of Rydberg atoms interacting with their surroundings. An important question they are interested in is the separation of the system from the environment which is treated as a bath, so back-reactions on the system are neglected. Authors give some results for the transition rates of Rydberg atoms. Instead of considering purequantum states, due to the interaction with the radiation field being always present, transitions to other states occur which result in a finite lifetime of the energy eigenstates. This effect is well known as natural line broadening, it generally demands the quantum description of the radiation field. If the generation of charged particles in a gas of Ryberg atoms can be suppressed, the strongest interaction is the resonant dipole-dipole multi-particle interaction resulting from the huge » n2 dipole matrixelement of the Rydberg atoms. The influence of this 1=R3 (R being the distance between two atoms) interaction on the dephasing of a two-atom superposition state has recently been studied. Even more effective in destroying the phase of a quantum state are Coulomb collisions. Thus,in Rydberg plasmas where Rydberg atoms are immersed into a plasma consisting of electrons and ions with density ne, ni, respectively, the transition rates coincide with the results for dephasing of electron wave packets in mesoscopic physics as long as the interaction is weak. For strong

Chapter 3 - The experimental study of transport in tokamaks has revealed a number of *strange phenomena*, apparently related to the fact that the plasma profiles are close to some critical threshold: Bohm scaling of the energy confinement time, stiff profiles, on-axis/off-axis heating paradoxes, superdiffusive pulse propagation, and others. These observations have cast some doubt on the appropriateness of the current description of transport, based on diffusivities and conductivities.In particular, the extrapolation of analysis results to larger devices may depend critically on this issue, with possibly important consequences for the design of next-generation tokamaks such as ITER. In this paper, authors will argue that the modelling of these strange phenomena requires a generalization of the usual diffusive transport paradigm on a fundamental level. In order to do so, recall that typical particle diffusion is the macroscopic consequence of the microscopic random motion of individual particles, described by a probability distribution function (pdf) with finite moments. This fact, combined with the central limit theorem that asserts that any sum of random variables obeying pdfs with finite moment converges to a Gaussian, yields the standard diffusive picture. However, Gaussians are not the only limit distributions produced by the central limit theorem: the larger family of Levy distributions is obtained when pdfs with divergent moments are considered instead. To illustrate these ideas, they will discuss the construction of several simple models in which the random particle motion is described by Levy distributions, governed by a critical gradient. The evolution equation for the particle density can then obtained in the form of a Master Equation or, by taking the fluid limit, of a Fractional Differential Equation. In spite of their simplicity, these models naturally produce the rapid propagation of pulses, Bohm-like scaling, and even on-axis peaking of the density in the presence of off-axis sources, similar to what is observed in actual tokamaks and stellarators. More importantly, this new approach shows that diffusivities and conductivities measured in relatively small devices may lead to error when extrapolated directly to predict transport in larger devices, and teaches us which physical quantities should be used instead to ensure that transport is properly scaled.

Chapter 4 - A rigorous study on edge plasma in the anchor cell of the GAMMA 10 tandem mirror is carried out for the first time using Langmuir probes, calorimeters, conducting plates (anchor plate (AP)), and H_α detectors. Computational studies on the magnetic filed configuration (MFC) of the GAMMA 10 and neutral transport phenomena are carried out to understand the experimental results. Probe current asymmetry is found in the minor axis direction of the elliptic flux tube at the non-axisymmetric MFC region. Measurements by calorimeters and APs support this plasma asymmetry. Some impurity deposited areas are observed on the APs and the pattern of these areas is consistent with the plasma asymmetry. Formation of comparatively high density, but low temperature plasma in the anchor cells is observed, especially during axial confinement. Significant effect of the floated AP on the GAMMA 10 plasma parameters is observed. Using three dimensional Monte-Carlo code, a modeling of neutral transport is successfully performed in the GAMMA 10 anchor cell with non-axisymmetric MFC, and the behavior of neutrals in this region is investigated on the basis of the experimental data of H_α intensities. The computational result well predicts the experimental result. Ambient neutral pressure control experiment with modulating gas fueling rate is performed and significant improvement of anchor plasma parameters is obtained. Possible explanations of these observations are given in detailed and adverse effects of no-axisymmetric MFC on GAMMA 10 plasma parameters are also pointed out. Finally, possible way to optimize the non-axisymmetric MFC of the anchor cell is discussed in part A of this chapter. Part B is concerned with the results of theoretical investigation on the properties and excitation of low frequency electrostatic dust modes, e.g., dust-acoustic (DA) and dust-lower-hybrid (DLH) waves, in a divertor plasma environment using the fluid model. In this study, dust grain charge is considered as a dynamical variable in streaming magnetized dusty plasmas with a back ground of neutral atoms/molecules. Dust charge fluctuation, collisional, and streaming effects on DA and DLH modes are discussed. Charging of dust grains is also addressed.

Chapter 5 - In the large helical device, LHD, material probe study has been conducted since the first experimental campaign in Mar. 1998. Material probes have been installed at the inner walls along the poloidal direction from the first experimental campaign, and both at the inner wall along the poloidal direction and at the first wall along the toroidal direction from the 4th experimental campaign. After each campaign, the surface morphology, the impurity deposition and the gas retention were examined by using surface analysis techniques in order to clarify the plasma surface interactions and the degree of wall cleaning. In the first experimental campaign, the iron oxide layer at surface was observed to be thick. However, in the 2nd campaign, the entire wall was thoroughly cleaned by glow discharge conditioning and the increase of main discharge shots. From the 3rd campaigns, graphite tiles were installed over the entire divertor strike region, and then the wall condition was significantly changed compared with the case of a stainless steel wall. It was seen that graphite tiles in the divertor were eroded mainly during main discharges, and the stainless steel first wall mainly during glow discharges. The eroded carbon during main discharges was deposited on the entire wall. The fraction of carbon coverage in the first wall was approximately 60%. The deposition thickness of carbon was large at the wall far from the plasma. The reduction of metal impurities in the plasma was observed, which corresponds to the carbonized wall. Since the entire wall was carbonized, the amount of retained discharge gases such as H and He became large. In particular, the helium retention was large at the position close to the anodes used for helium glow discharge cleanings. One characteristic of the LHD wall is a large retention of helium since the wall temperature is limited to below 368 K. From the 5th experimental campaigns, boronization was several times conducted in each campaign to control the gas retention in the wall. The fraction of boronized wall to the entire wall was 20-30%. A large reduction of oxygen impurity level in the plasma was observed after the boronization. This result shows that the oxygen was well trapped in the boron layer even if the coverage of boron was small. The results on the material probes have been referred for the next experimental campaign, and then the plasma confinement was significantly improved. At present, the plasma density limit exceeded $10^{20}m^{-3}$, both electron and ion temperatures 10keV and the average beta 4%. These parameters are comparable with those of existing large tokamaks such as JT-60U and JET.

Chapter 6 - Nonlinear interaction of an intense electromagnetic (EM) beam with relativistically hot electron-positron plasma is investigated by invoking the variational principle and numerical simulation, resting on the model of generalized nonlinear Schr"odinger equation with saturating nonlinearity. The present analysis shows the dynamical properties including the possibilities of trapping and wave-breaking of EM beams. These properties of EM beams may give a significant clue for the gamma-ray burst.

Chapter 7 - There are three conventional techniques used to fuel fusion devices: gas puffing, ice pellet injection and neutral beam injection. Gas puffing is the simplest fuelling tool and it is generally used in all devices to establish primary plasma and control the plasma density by feedback, but the fuelling efficiency is quite low, in the range of 5-25 %. On the other hand, a pellet injected from inside the magnetic axis from the inner wall leads to stronger central mass deposition and thus yields deeper and more efficient fuelling. However, this injection system meets complex problems related to producing and launching as well as transporting ice pellets. The main goal of neutral beam injection is the heating of the plasma, but not fuelling the large fusion plasmas. Supersonic molecular beam injection (SMBI) was first proposed and demonstrated on the HL-1 tokamak, was successfully developed and used on the HL-1M tokamak, and was then applied on the HT-7 superconducting tokamak, the Tore Supra superconducting tokamak, the W7-AS stellarator and the ASDEX Upgrade tokamak. SMBI can enhance the penetration depth and fuelling efficiency in the previous devices. With the new fuelling method, high densities of 8.2×10^{19} m^{-3} and 6×10^{19} m^{-3} were obtained for HL-1M and HT-7, respectively. A stair-shaped density increment was obtained with high-pressure multi-pulse SMBI that was just like the density evolution behavior during multi-pellets injection. A pneumatic pulsed SMB injector was developed in CEA/DSM/DRFC at Cadarache, which can work in presence of strong magnetic field and greatly increase fuelling efficiency in Tore Supra for 3 times than that of conventional gas puff. Considering the relatively high temperature of edge plasma for the large tokamak with divertor configuration compared with the limiter one, a cluster jet injection (CJI), which is like the micro-pellet injection, will be beneficial to deeper injection and higher fuelling efficiency for both of the SMBI and gas puffing. The experiment on cryogenically cooled high-pressure hydrogen cluster jet injection into the HL-2A plasma was carried out and the fuelling effects were distinctly better than that of the room temperature one. This technique may be a candidate for the fuelling of the International Thermonuclear Experimental Reactor (ITER).

Chapter 8 - Physics of fast wave current drive and electron heating at a frequency range of ion cyclotron resonance have been investigated in fusion devices. The theoretical studies predicted the fine structure of electromagnetic field of fast waves in fusion plasmas. However, behaviors of fast waves in plasmas can be estimated from macroscopic experimental results, for example, the reduction of toroidal loop voltage and the motional stark effect with the poloidal magnetic field. In this chapter, the new method for diagnosing electromagnetic field pattern of fast waves in toroidal plasmas is introduced. The fluctuation of ponderomotive potential at a beat wave frequency between the two frequency fast waves produces the actual potential fluctuation via electron and ion motions in fusion plasmas. The amplitude of potential fluctuations are proportional to the square of electric field strength of fast waves. Therefore, the amplitude pattern of potential fluctuations at a beat wave frequency indicates the field pattern of fast wave in plasmas. The potential fluctuation patterns in high temperature plasma can be detected with a heavy ion beam probe (HIBP), so that the pattern of electromagnetic field of fast wave can be estimated from the data of HIBP with the beat wave technique. The potential fluctuation pattern at the beat wave frequency (90 kHz) has been measured with HIBP system during the fast wave pulses at a frequency of 200 MHz in JFT-2M, which is a middle size non-circular tokamak. The measured levels of potential fluctuations decrease with increasing an electron temperature, consistent with the improvement of wave absorption. The measured potential fluctuation levels are similar to the theoretical predictions. The feasibility of this method in large fusion devices depends on the wave absorption and the sensitivity of HIBP.

Versions of these chapters were also published in *Journal of Magnetohydrodynamics, Plasma and Space Reserach,* Volume 12, Numbers 1-4, published by Nova Science Publishers, Inc. They were submitted for appropriate modifications in an effort to encourage wider dissemination of research.

In: Advances in Plasma Physics Research, Volume 7
Editor: Francois Gerard

ISBN: 978-1-61122-983-7
© 2011 Nova Science Publishers, Inc.

Chapter 1

THE DYNAMICS OF THE PLASMASPHERE

Viviane Pierrard[*]

Belgian Institute for Space Aeronomy, 3 av. Circulaire, B-1180 Brussels, Belgium

ABSTRACT

The plasmasphere is the high altitude extension of the ionosphere. To ensure accuracy and reliability of communications, navigation and other satellites stationed in this region, there is a considerable interest to understand the plasmaspheric environment and its dependence on external parameters.

Dynamical simulations have been developed at the Belgian Institute for Space Aeronomy to simulate the deformation of the plasmasphere during geomagnetic substorms and other variations in the level of geomagnetic activity. These simulations are based on the mechanism of plasma instability and on the E5D empirical electric field model. The results of these simulations are compared with satellite observations of the plasmasphere. Comparisons with observations of the IMAGE spacecraft launched in March 2000 are especially useful to understand the dynamics of the plasmasphere since the satellite provides the first global comprehensive images of the Earth's plasmasphere. The EUV (Extreme UltraViolet) instrument gives intensity maps of the 30,4 nm emissions of Helium ions integrated along the line of sight. It revealed the global behavior of the plasmasphere and its dynamics influenced by the solar wind and interplanetary magnetic field variations. New structures in the plasmasphere such as shoulders, channels, notches and the formation of plumes were discovered. The four Cluster spacecraft launched in 2000 also provide useful data to compare with the simulations.

INTRODUCTION

The plasmasphere is filled by cold (a few eV or less) plasma originating from the ionosphere distributed along geomagnetic field lines and co-rotating with the Earth. This toroidal region encircles the Earth at geomagnetic latitudes typically less than about 65

---

[*] E-mail address: viviane.pierrard@aeronomie.be; Tel: (+32) 2 3730418; Fax: (+32) 2 3748423

degree. This region where the number density of the particles is in the range of $10\text{-}10^4$ particles/cm^3 was discovered at the end of the fifties with the pioneering ground-based whistler wave investigations and confirmed with the first in situ observations of plasma detectors on board the Soviet LUNIK 1 and 2 spacecraft launched in 1959. Besides ground-based whistler technique and direct measurements of electron and ion number density, other techniques to observe the plasmasphere were then developed, based on plasma wave experiments and remote sensing by optical means.

The observations showed that the number density of the particles decreases sometimes sharply at the limit of the plasmasphere, called the plasmapause. This discontinuity crosses the geomagnetic equatorial plane at radial distances ranging from 2 Re to 7 Re depending on the geomagnetic activity level. The plasmapause is closer to the Earth during geomagnetic substorms.

After four decades of study by ground-based and in situ measurements, many of the details of the plasma distribution remain unexplained and the processes in the inner magnetosphere poorly understood [Carpenter, 1995].

New progress in the plasmaspheric field of research is provided by the recent observations of EUV on board IMAGE [Sandel et al., 2003]. IMAGE provides indeed the first global views of the plasmasphere by imaging the distribution of He$^+$ in its 30.4 nm resonance line. The details of the plasmaspheric dynamics are then revealed: new structures such as shoulders, channels, notches and the formation of plumes, appear frequently in the observations. These are keys to understanding the ways that electric fields affect the plasma distribution and the mechanisms influencing the formation of the plasmapause.

Different processes have been proposed for the formation of the plasmapause [see Lemaire and Gringauz, 1998 for a review]. The role of the convection electric field and of the instability mechanism has been studied since 1974 at the Belgian Institute for Space Aeronomy [Lemaire, 1974, 1985]. Recently, dynamical simulations based on this instability mechanism have been developed [Pierrard and Lemaire, 2004] and compared with new observations of the plasmasphere by AURORAL PROBE, CLUSTER, and IMAGE [Bezrukikh et al., 2003; Pierrard, 2004; Dandouras et al. 2005; Pierrard and Cabrera, 2005; 2006].

AURORAL-PROBE shows clearly that the plasmasphere is eroded during geomagnetic substorms and that the new plasmapause is not the result of compression. Indeed, the number density inside the plasmasphere is not increased after the geomagnetic event.

This spacecraft also measured the temperature in the plasmasphere and shows large difference between dayside and nightside.

CLUSTER launched in 2000 revealed new density structures in the plasmasphere and the plasmapause [Darrouzet et al., 2004]. These four spacecraft flight on a polar orbit in a tetrahedral configuration to study the magnetosphere and its environment in three dimensions. The inter-satellite distance has been varied from 100 km to several thousand kilometers. The satellites penetrate periodically (57 h) inside the plasmasphere when they are at the perigee at 4 Re. The instrument WHISPER measures number density profiles of the electron by operating a relaxation sounder. CLUSTER crosses the plasmapause only when the limit of the plasmasphere is further from the Earth than 4 Rs, thus during quiet geomagnetic activity. The comparisons between the global view of IMAGE and the density profiles of CLUSTER gave interesting results about the plume formation and the ionic structures among others.

In the present chapter of this book, new study cases of the dynamical simulations based on the instability mechanism are presented and compared with the observations of EUV/IMAGE. These typical cases show the extended plasmasphere during quiet periods and the formation and evolution of plumes during geomagnetic substorms. The advances in plasmaspheric research as revealed by EUV and CLUSTER are also summarized.

PLASMAPAUSE FORMATION

The dynamical simulations presented in the present paper use a given Kp-dependent magnetospheric electric field model. This electric field is the E5D model determined from dynamical proton and electron spectra measured on board the geostationary satellites ATS-5 and 6 [McIlwain, 1986]. The Kp index observed during the analyzed period of time and the previous 24 hours is the only input of the time-dependent model. The convection electric field intensity is the largest in the post-midnight sector. Due to the increased convection velocity during periods of increased geomagnetic activity, the plasmasphere is peeled off in this region, according to the physical mechanism of plasma instability used in the numerical simulations. The instability mechanism and the simulation process are described in detail in Pierrard and Lemaire [2004]. Since plasma is detached in the post-midnight sector due to instability during geomagnetic substorms, the plasmapause is then closer to the Earth in this sector. Moreover, due to the differential rotation, a plume is formed later in the afternoon and dusk region.

The model predictions are compared with EUV observations from IMAGE. The instrument consists of three wide-field cameras that are tuned to the 30.4 nm resonance line of sunlight. The observations are intensity maps of these emissions of Helium ions integrated along the line of sight. They are projected in the geomagnetic equatorial plane to have the same view over the pole as in the simulations. In the Figures, sunlight is incident from the left. The plasmapause is assumed to be the sharp edge where the brightness of 30.4 nm He^+ emissions drops drastically. To better visualize the plasmapause, a white line is drawn corresponding to a threshold of a percentage of the maximum intensity of the image, where the intensity is the logarithm of the luminosity.

JUNE 18, 2001

First, let us show the example of June 18, 2001. During this day, Kp increased up to 5+ and remained high during several hours. The upper panel of Fig. 1a shows the geomagnetic activity index Kp observed from June 16 to June 18, 2001. The Dst index that becomes negative during the geomagnetic substorm event is also illustrated.

The lower panel shows the result of the simulation on June 18, 2001 at 13h00 UT. Kp is then observed to be maximum. The colored dots represent successive positions of test plasma elements released every 10 minutes at 02:00 MLT in the post-midnight sector. These dots identify the geomagnetic equatorial position of the plasmapause, as predicted by the mechanism of interchange motion. The position of the plasmapause depends on the evolution of the geomagnetic index Kp during the previous hours, which controls the intensity of the convection electric field E5D. The different colors of the elements correspond to the epoch of their release. The dotted circles correspond to L=1, 2, 3, 4 and 5.

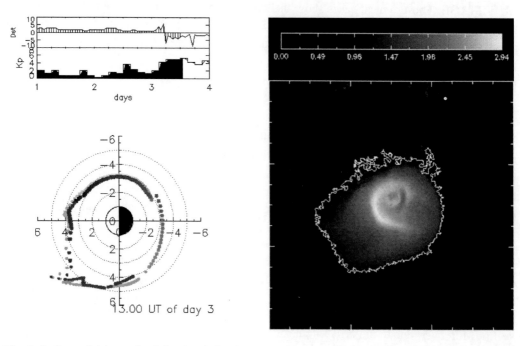

Fig. 1: Left panel (a): result of the simulation based on the instability mechanism, the E5D model and the value of Kp on 18 June 2001, 13h00. The dots indicate the position of plasmapause in the geomagnetic equatorial plane. The indexes Dst and Kp observed during the previous days are also displayed. The dotted circles correspond to L=1, 2, 3, 4 and 5.

Right panel (b): EUV observations on 18 June 2001 at 13h06, projected in the geomagnetic equatorial plane. The white line corresponds to 35% of the maximum intensity of the image and permits to visualize the plasmapause. The limit of the black square corresponds to L = 8.

The simulation predicts the development of a so-called plume in the afternoon region of the plasmasphere. Such plumes appear especially during periods of geomagnetic substorm events, i.e., after an increase of Kp.

The result of the simulation is compared with the plasmasphere observed by EUV/IMAGE on June 18, 2001 at 13h06 illustrated in Fig. 1b. The plasmasphere is viewed from a point of view about the North pole and projected in the geomagnetic equatorial plane. The edges of the plane extend up to 8 Rs. The observations of EUV/IMAGE show indeed the presence of a plume in this sector. The position of the plasmapause is about 3 Re in the midnight LT sector, i.e. close to the Earth, as in the simulations.

The white line corresponds to 35% of the maximum luminosity and illustrates the position of the plasmapause as observed by EUV. A smooth decrease in density is observed in the dayside sector. The dayside position the plasmapause would have been found at lower radial distances in the dayside sector if a higher threshold had been chosen. On the contrary, the gradient in density is sharper in the night sector. A sharp plasmapause knee is predicted by the mechanism of instability in this sector, since it is the region where the plasmasphere is peeled off during geomagnetic substorms.

Plumes are often observed by EUV/IMAGE and follows even moderate increases of Kp (above 3^+-4). Unfortunately, there are no observations in the present case study during the hours preceding its formation. But the EUV/IMAGE observations, which have generally a typical duration of 7 hours (with a time resolution of 10 minutes) out of each 14-hour orbit, allow us to follow the evolution of the plume during the next hours.

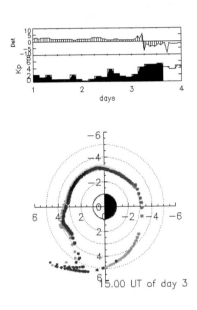
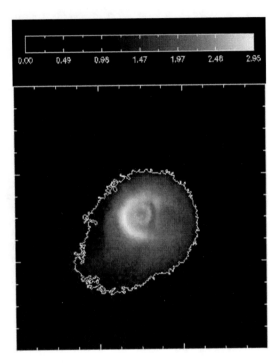

Fig. 2: Same as Fig. 1 for 18 June 2001 at 15h00 (left panel (a), simulation) and 14h59 UT (right panel (b), EUV observation).

Fig. 2 a and b present the plasmasphere two hours later than in Fig. 1, respectively as predicted by the model at 15h00 and as observed by EUV at 14h59. Since Kp remains high, the plasmapause remains very close to the Earth in the post-midnight sector. The base of the plume corotates with the Earth. But its top located at a large radial distance has a lower velocity, so that the plume becomes more and more bent and wraps. Moreover, it becomes thinner with time since the inner edge rotates faster than the outer edge. This evolution of the plume seems quite general, since it was observed also in other sequences [Spasojevic et al., 2003].

The evolution of the plume continues during the following hours and is illustrated in the next Figures. Fig. 3a shows the result of the simulation at 18h00 and 4a at 20h00. The corresponding EUV/IMAGE observations are displayed on Fig. 3b (18h03) and Fig. 4b (20h06). The comparison shows that the simulations reproduce quite well the formation of the plume due to the Kp increase and its evolution with time. This is confirmed also by other study cases obtained during geomagnetic substorms [Pierrard and Cabrera, 2005].

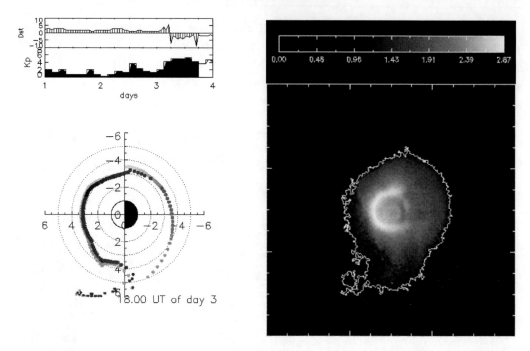

Fig. 3: Same as Fig. 1 for 18 June 2001, 18h00 (left panel (a), simulation) and 18h03 UT (right panel (b), EUV observation).

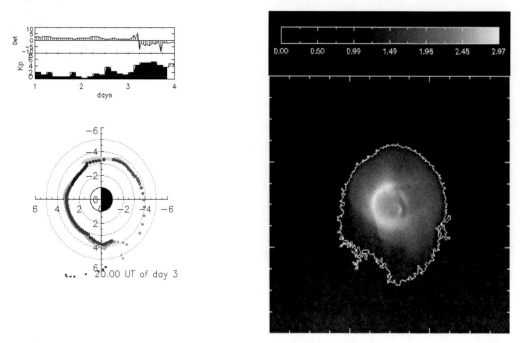

Fig. 4: Same as Fig. 1 for 18 June 2001, 20h00 (left panel (a), simulation) and 20h06 UT (right panel (b), EUV observation).

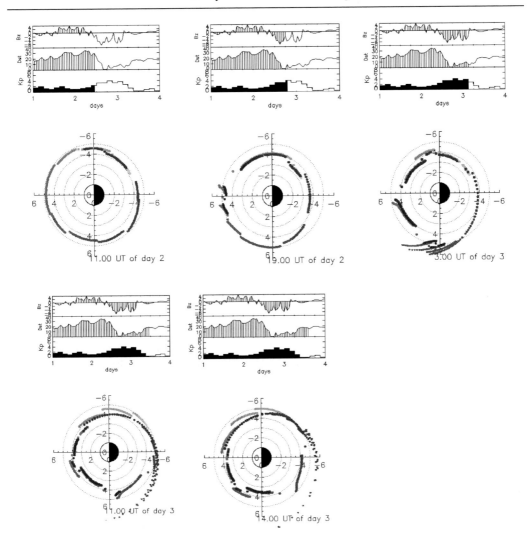

Fig. 5: Sequence showing the result of the simulation based on the instability mechanism, the E5D model and the value of Kp on 26 June, 2001, 11h00 (a); 19h00 (b); 27 June 2001, 3h00 (c); 11h00 (d) and 14h00 (e). The plasmapause in the geomagnetic equatorial plane corresponds to the dots. The indexes Bz, Dst and Kp observed during the previous and following days are also displayed. The dotted circles correspond to L=1, 2, 3, 4 and 5.

26-27 JUNE, 2001

Another interesting study case corresponds to the relatively weak geomagnetic substorm of 26-27 June 2001. Kp increased gradually up to 4 and then gradually decreased. This event is weak but well isolated. In the upper panels of Fig. 5, the Kp index is illustrated, as well as Dst and Bz, the z component of the interplanetary magnetic field. Note that the geomagnetic substorm corresponds to an increase of Kp, to a decrease of Dst and to Bz becoming southward.

This sequence of IMAGE had been analyzed by Spasojevic et al [2003]. It is interesting because it is possible to follow the development of the plume and its evolution during two days. The EUV/IMAGE observations are interrupted during some hours but observations are available on June 26 from 4h31 to 14h35 and from 18h39 to 1h23 and from 9h39 to 18h31 on June 27. The rotation velocity is observed to be 80% of co-rotation during this period. This observed co-rotation lag has been accounted for in the model.

The result of the simulation is displayed on the sequence presented in Fig. 5.

The first panel corresponds to June 26, 2001 at 11h00. The plasmapause is then located between 4 and 5 Re following the LT sectors and is quite circular. Indeed, before the geomagnetic substorm, Kp was small and quite constant. Its irregular shape is due to the small variations of Kp during the previous 24 h. Note that the plasmapause is a little bit closer to the Earth in the post-midnight sector. This is due to the small increase of Kp at the beginning of the substorm. Indeed, the interchange mechanism predicts a plasmapause closer to the Earth in the post-midnight sector when Kp increases [Pierrard and Lemaire, 2004].

A little bit later, a plume begins to form in the morning sector. This is observed in the second panel showing the result of the simulation at 19h00. Later, the plume becomes thinner and more elongated. It rotates and is located in the afternoon sector at 3h00 (third panel) on 27 June 2001, in the dusk sector at 11h00 (fourth panel) and enters in the post-midnight sector at 14h00 (fifth panel).

Let us now compare with observations of EUV/IMAGE obtained during the same period of time. Global EUV images of the plasmasphere are available during three intervals of time on June 26 and 27, when the satellite was close to its apogee.

We use the threshold 0.45 of the maximum intensity in this sequence to visualize the plasmapause by a white line. The threshold cannot be chosen to be lower during this observation period, otherwise the white line corresponds in some cases to the limit of the camera view. This happens especially when the satellite is far from the apogee. The relative intensity threshold procedure is automatic and provides slightly different plasmapause positions than those obtained by the manual procedure used by Spasojevic et al. [2003]. The plasmapause is located at larger radial distances when a lower threshold is chosen. Note nevertheless that the modification of the threshold does not modify the position of the plasmapause in the postmidnight sector, since the gradients are generally sharper in this sector. White circles are also displayed on the observations to better evaluate the radial distance of the observed plasmapause: they correspond to L=1, 2, 4, 6 and 8.

The sequence shows indeed the formation of a plume, as predicted by the model. The plume begins its formation in the morning sector like in the simulations (see Fig. 6b). The formation of the plume is generally thought to be located in the afternoon sector, but one can see that in the present case, a small bulge is already visible in the morning sector. On Fig. 6c, d and e, the plume position corresponds quite well to what is obtained in the simulation. Nevertheless, the position of the plasmapause is generally located further in the simulations than what is observed. This effect can be due to the large value of the threshold chosen in the present case. Note also that the observations show an erosion at the western edge of the plume, with crenulations and irregularities. The plasmasphere is eroded in the dawn sector, while the plasmapause is more circular in the simulations.

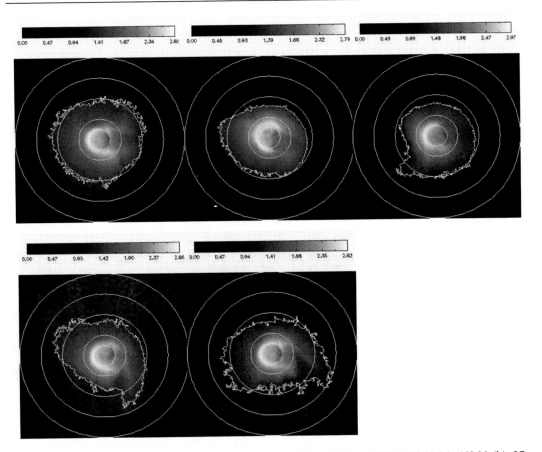

Fig. 6: Sequence showing the observations of EUV/IMAGE on 26 June, 2001, 11h00 (a); 19h00 (b); 27 June 2001, 3h00 (c); 11h00 (d) and 14h00 (e). The white line corresponds to a threshold of 45% of the maximum intensity and corresponds to the plasmapause in the geomagnetic equatorial plane. The white circles correspond to L=1, 2, 4, 6 and 8.

Other comparisons between EUV/IMAGE observations and results of simulation are also presented in Pierrard and Cabrera [2005], like for instance 9-10 June 2001 and 25-26 June, 2000. These events show an extended plasmasphere during periods of low geomagnetic activity level and the formation of a plume during geomagnetic substorms. Plumes begin to form in the morning LT sector and are well developed in the afternoon LT sector during geomagnetic substorms and storms. Subsequently, they tend to co-rotate with the Earth when the geomagnetic activity decreases. After a sharp decrease of Kp, the formation of shoulders has also been observed. The global evolution of the plume and the radial distance of the plasmapause in the different LT sectors are generally well reproduced by the simulations. But of course, the model has inherent limitations and is not expected to reproduce all observed features. The empiric 3-hourly Kp dependent quasi-static electric field model E5D is a global scale average: it can be quite different from the actual E-field during some periods of time.

IMAGE AND CLUSTER REVELATIONS ABOUT THE PLASMASPHERE

The main observed characteristics and theories developed about the plasmasphere before 1998 are well summarized in Lemaire and Gringauz [1998]. Since then, many new results were obtained with the American IMAGE and the European CLUSTER missions, launched in 2000.

Scientific results about the whole magnetosphere were obtained by the IMAGE mission [Burch, 2003]. During the first two years of the IMAGE mission, most EUV observations have been from high latitudes with a favorable viewpoint for studying the distribution of plasma in the equatorial plane. For the first time, it was possible to observe the plasmapause in different MLT. During the extended mission of IMAGE between 2003-2005, the imaging perspective has changed to lower latitudes and the magnetospheric environment undergoes a transition from solar maximum to solar minimum.

Global imaging of the plasma distribution inside the plasmasphere led to new types of observations and of investigations summarized in this section.

1. Comparisons between EUV images and in situ measurements (RIP aboard IMAGE for instance) allow the determination of the particles number density inside the plasmasphere and show that the sharp outer boundary in EUV images (detecting He^+ ions) corresponds indeed to the plasmapause [Goldstein et al., 2003c].

2. Formation of plumes is a common consequence of increased convection. As they form, the connection is in the afternoon or dusk sector and the plumes extend sunward. The base of the plumes generally co-rotates [Spasojevic et al., 2003].

3. "Shoulders", asymmetric bulges in the plasmapause, were unexpected features of the plasmasphere seen for the first time in EUV images [Burch et al., 2001].

4. "Notches" were also discovered by EUV: they are localized density cavities in the plasmapause.

5. In times of increasing magnetospheric convection, the plasmasphere shrinks rapidly. The observations show that the plasmasphere is eroded, not compressed. Erosion occurs over the antisunward hemisphere, but not uniformly. The plasmapause moves at different times and at different rates at different local times. The plasmapause can move fast inward, for instance by almost 2 Re in less than 3 hours on 10 July 2000 [Goldstein et al., 2003a].

6. When convection returns to normal, flux tubes depleted by convection refill from the ionosphere. This process is slow but can be studied with EUV observations. During prolonged periods of geomagnetically quiet conditions, the plasmasphere extends to L > 5. The plasmaspheric refilling has been observed to be completed in less than 28 h (2 IMAGE orbital periods), which is less than theoretical expectations (2-3 days) [Reinisch et al., 2003].

7. "Channels" are regions of lower density extending roughly in the azimuthal direction. They usually appear first in the pre-midnight sector and allow the measurement of the rotation velocity. Some channels are formed by plume wrapping: the inner part of a plume nearly co-rotates while the outer part remains nearly fixed in local time.

8. Low-density regions (notches, channels) can serve as natural markers to track the angular velocity of the plasma over a range of L values. Some notches can persist for periods as long as 60 hours. Cold plasma in the range 2<L<4 most frequently rotates at a rate 85-90 % of the co-rotation. But it may move at the co-rotation velocity for a time and then begin to lag.

The plasmasphere corotation lag is observed to be linked to a corresponding corotation lag in the ionosphere [Burch et al., 2004].

9. A comparison of how well three different electric field models can predict the plasmapause position as observed by EUV during the magnetic storm of 17 April 2002 is provided by Liemohn et al. [2004].

10. The multiple instruments on IMAGE give a comprehensive image of the entire ionosphere-plasmasphere-magnetosphere coupling [Goldstein et al., 2005]. Links between plasmaspheric plumes and sub-auroral proton arcs have been detected [Spasojevic et al., 2004]. There is an increasing interest on the role of the sub-auroral polarization stream (SAPS) in plume formation and in the sunward transport of plasma from dusk [Foster and Vo, 2002; Goldstein et al., 2003b].

Moreover, the four CLUSTER spacecraft launched in 2000 revealed also many important features in the magnetosphere. The local measurements of 4 spacecraft, combined with correlating global images of EUV, are useful to deconvolve spatial from temporal effects. The main new results concerning the plasmasphere obtained with these exceptional data are summarized in the next section:

1. The instrument WHISPER measures the electron number density profiles along the orbit of the spacecraft. Two different types of plasmasphere crossings are observed: a) smooth density variations with a sharp and well-defined knee corresponding to the plasmapause and b) small-scale density structures observed in the plasmasphere and the plasmapause. Due to the time resolution of 2 seconds, these thin structures and density irregularities (density variations of a factor 2 on 10 km) were for the first time observed by CLUSTER [Darrouzet et al., 2004].

2. Apparently detached plasma elements observed by both instruments are in many cases plasmaspheric plumes. The density values measured during the detached plasmasphere observations are by about an order of magnitude lower than the ones measured in the main plasmasphere. The thin structures and the density decrease are consistent with the density reduction for the detached plasma shells predicted by the peeling-off models due to plasma instability. These observations show that plumes have large latitudinal extent.

3. The Cluster Ion Spectrometry (CIS) instruments provide full three-dimensional distributions for the main ion species (H^+, He^+, O^+). It shows that H^+ and He^+ present mostly similar profiles with the He^+ densities being lower by a factor of ~15. On the contrary, O^+ is not observed as trapped plasmaspheric population at CLUSTER orbit altitude (R > 4 Re) [Dandouras et al., 2005].

4. Low-energy ($E < 25$ eV) O^+ is observed only as upwelling ions, escaping from the ionosphere along auroral field lines. O^{++} can be also observed in these upwelling ion populations.

5. In the vicinity of the plasmapause, around the geomagnetic equator, banded hiss-like electromagnetic emissions (2 to 10 kHz) are often observed [Masson et al., 2004].

6. EFW (Electric Field and Wave) instrument on Cluster give some simultaneous electric field and density observations [Gustafsson et al., 2001].

CONCLUSION

We presented here two detailed studies of the global dynamics of the plasmasphere during geomagnetically disturbed periods. A larger set of cases was studied and other sequences were presented in Pierrard and Cabrera [2005] and Dandouras et al. [2005].

From the direct comparison between the EUV observations from IMAGE and the numerical simulations of the plasmapause position determined from the value of Kp using the instability mechanism and the E5D electric field model, it appears that the formation of the plumes and their evolution with time are globally well reproduced. Plumes are formed after a moderate increase of Kp and rotate then with the plasmasphere during the quieting period following the enhanced magnetic activity. The observed shoulders are produced just at the end of magnetic substorms or storms and generally develop one or two hours before a sharp decrease of Kp. The plasmapause position obtained with the simulations corresponds in average to the position observed by EUV: it is not simply determined by the instantaneous Kp value, but it depends also on its past history.

The plasmasphere is highly structured and complex. Reproducible features evolve clearly during disturbed periods, but some differences appear also. The new observations of EUV/IMAGE will help to improve the theory for the formation of the plasmapause and the understanding of the new structures recently discovered in the plasmasphere.

ACKNOWLEDGMENTS

V. Pierrard thanks Dr J. Cabrera (Center for Space Radiations, UCL) for the image processing and Prof. J. Lemaire and L. Dricot for their contributions in the development of the dynamical simulations. She thanks also Dr. B. Sandel, Lunar and Planetary Laboratory, University of Arizona, Tucson, USA and Dr. D.L. Gallagher, NASA, National Space Science and Technology Center, Huntsville, USA for the access to the EUV observations of the satellite IMAGE.

This project is financially supported by the Belgian Federal Science Policy Office (PPS).

REFERENCES

Bezrukikh V. V., Kotova G. A., Lezhen L. A., Lemaire J., Pierrard V. and Venediktov Yu. I. (2003). Dynamics of temperature and density of cold protons of the Earth's plasmasphere measured by the Auroral Probe/Alpha-3 experiment data during geomagnetic disturbances. *Cosmic Research*, Vol. 41, No. 4, 392-402.

Burch J. L. (2003). The first two years of IMAGE. *Space Sci. Rev.*, 109, 1-24.

Burch J. L., Goldstein J. and Sandel B. R. (2004). Cause of plasmasphere corotation lag. *Geophys. Res. Lett.*, 31, L05802, doi:10.1029/2003GL019164.

Burch J. L., Mende S. B., Mitchell D. G., Moore T. E., Pollock C. J., Reinisch B. W., Sandel B. R., Fuselier S. A., Gallagher D. L., Green J. L., Perez J. D., and Reiff. P. H. (2001). Views of Earth's Magnetosphere with the IMAGE Satellite. *Science*, 291, 619-624.

Carpenter D. L. (1995). Earth's plasmasphere awaits rediscovery. *EOS Trans. Am. Geophys. U.*, 76(9), 89.

Dandouras I., Pierrard V., Goldstein J., Vallat C., Parks G. K., Rème H., Gouillart C., Sevestre F., McCarthy M., Kistler L. M., Klecker B., Korth A., Bavassano-Cattaneo M. B., Escoubet Ph., and Masson A. (2005). Multipoint observations of ionic structures in the Plasmasphere by CLUSTER-CIS and comparisons with IMAGE-EUV observations and with Model Simulations. Inner Magnetosphere Interactions: New Perspectives from Imaging, *AGU Geophysical Monograph*, 159, 23-54, doi: 10.1029/2004BK000104.

Darrouzet F., Décréau P. M. E., De Keyser J., Masson A., Gallagher D. L., Santolik O., Sandel B. R., Trotignon J. G., Rauch J. L., Le Guirriec E., Canu P., Moullard O., Sedgemore F., André A. and Lemaire J. F. (2004) Density structures inside the plasmasphere: Cluster observations. *Annales Geophys.*, 22: 2577-2585.

Foster, J. C., and Vo H. B. (2002). Average characteristics and activity dependence of the subauroral polarization stream. *J. Geophys. Res.*, 107(A12), 1475, doi: 10.1029/2002JA009409.

Goldstein, J., Burch J. L., Sandel B. R., Mende S. B., son Brandt P. C., and Hairston M. R. (2005). Coupled response of the inner magnetosphere and ionosphere on 17 April 2002. *J. Geophys. Res.*, 110, A03205, doi:10.1029/ 2004JA010712.

Goldstein J., Sandel B. R., Forrester W. T., and Reiff P. H. (2003a). IMF-driven plasmasphere erosion of 10 July 2000. *Geophys. Res. Lett.*, 30, 3, 1146, 10.1029/ 2002GL016478.

Goldstein, J., Sandel, B. R., Hairston, M. R., and Reiff, P. H. (2003b). Control of plasmaspheric dynamics by both convection and sub-auroral polarization stream. *J. Geophys. Res.*, 30 (24), 2243, doi: 10.1029/2003GL018390.

Goldstein J., Spasojevic M., Reiff P. H., Sandel B. R., Forrester W. T., Gallagher D. L., and Reinisch.B. W. (2003c). Identifying the plasmapause in IMAGE EUV data using IMAGE RPI in situ steep density gradients. *J. Geophys. Res.*, 108, A4, 1147, doi:10.1029/2002JA009475.

Gustafsson G., André M., Carozzi T., Eriksson A. I., Falthammar C.-G., Grard R., Holmgren G., Holtet J. A., Ivchenko N., Karlsson T., Khotyaintsev Y., Klimov S., Laakso H., Lindqvist P.-A., Lybekk B., Marklund G., Mozer F., Mursula K., Pedersen A., Popielawska B., Savin S., Stasiewicz K., Tanskanen P., Vaivads A., and Wahlund J.-E. (2001). First results of electric field and density observations by Cluster EFW based on initial months of operation. *Annales Geophys.*, 19: 1219-1240.

Lemaire J. (1974). The 'Roche limit' of ionospheric plasma and the formation of the plasmapause. *Planet. Space Sci.*, 22, 757-766.

Lemaire J. F. (1985). *Frontiers of the plasmasphere* (Theoretical aspects). Université Catholique de Louvain, Faculté des Sciences, Editions Cabay, Louvain-La-Neuve, ISBN-2-87077-310-2.

Lemaire J. F. and Gringauz K. I. *The Earth's plasmasphere*; Cambridge University Press, United Kingdom, 1998, 350 p.

Liemohn M. W., Ridley, A. J., Gallagher, D. L., Ober, D. M. and Kozyra J. U. (2004). Dependence of plasmaspheric morphology on the electric field description during the recovery phase of the 17 April 2002 magnetic storm. *J. Geophys. Res.*, 109, A03209, doi:10.1029/2003JA010304.

Masson A., Inan U. S., Laakso H., Santolik O., and Décréau P. (2004). Cluster observations of mid-latitude hiss near the plasmapause. *Annales Geophys.*, 22:2565-2575.

McIlwain C. E. (1986). A Kp dependent equatorial electric field model, The Physics of Thermal plasma in the magnetosphere. *Adv. in Space Res.*, 6 (3), 187-197.

Pierrard V. (2004). La plasmasphère vue par IMAGE. *Ciel et Terre*, 120, 4, 114-117.

Pierrard V. and Cabrera J. (2005). Comparisons between EUV/IMAGE observations and numerical simulations of the plasmapause formation. *Annales Geophysicae,* 23, 7, 2635, SRef-ID: 1432-0576/ag/2005-23-2635.

Pierrard V. and Cabrera J. (2006). Dynamical simulations of plasmapause deformations. In Press in *Space Science Review.*

Pierrard V. and Lemaire J. (2004). Development of shoulders and plumes in the frame of the interchange instability mechanism for plasmapause formation. *Geophys. Res. Lett.*, 31, 5, L05809, 10.1029/2003GL018919.

Reinisch B. W., Huang X., Song P., Green J. L., Fung S. F., Vasyliunas V. M., Gallagher D. L., and Sandel B. R. (2003). Plasmaspheric mass loss and refilling as a result of a magnetic storm. *J. Geophys. Res.*, 10.1029/2003JA009948, 1-11.

Sandel B. R., Goldstein J., Gallagher D. L.and Spasojevic M. (2003). Extreme ultraviolet imager observations of the structure and dynamics of the plasmasphere. *Space Sci. Rev.*, 109, 25-46.

Spasojevic M., Frey H. U., Thomsen M. F., Fuselier S. A, Gary S. P., Sandel B. R., and Inan U. S. (2004). The link between a detached subauroral proton arc and a plasmaspheric plume. *Geophys. Res. Lett.*, 31, L04803, doi:10.1029/2003GL018389.

Spasojevic M., Goldstein J., Carpenter D. L., Inan U. S., Moldwin M. B., and Reinisch B. W. (2003). Global response of the plasmasphere to a geomagnetic disturbance. *J. Geophys. Res.*, 108, 1340.

In: Advances in Plasma Physics Research, Volume 7   ISBN: 978-1-61122-983-7
Editor: Francois Gerard   © 2011 Nova Science Publishers, Inc.

Chapter 2

Dephasing in Rydberg Plasmas

C. Gocke and G. Röpke

University of Rostock, Institut für Physik, Universitätsplatz 3
D-18051 Rostock, Germany

1. Introduction

In Coulomb systems, bound states can be formed which are obtained from the solution of the two-particle Schrödinger equation. Due to the interaction with the surroundings, transitions between these stationary states are possible so that a finite lifetime for the bound states is observed. In particular, Coulomb interaction as collisions with other charged particles as well as the interaction with the transverse part of the Maxwell field, leading to the emission and absorption of photons, can be considered using the method of thermodynamic Green functions, see, e.g., Ref. [1].

In a more general context, the finite lifetime of the bound states can be treated as the destruction of phase coherence due to the influence of the surrounding bath. On the other hand, decoherence and dephasing are concepts to characterize the damping of off-diagonal elements in the density operator [2]. At present, decoherence is an important concept to investigate open quantum systems [3], which are in contact with a bath so that irreversible time evolution is possible, see Ref. [4].

We will consider Rydberg atoms which are weakly bound electrons in highly excited atomic states. Measurements of the transition rates for Rydberg states with principal quantum number $n = 10 - 40$ due to Coulomb interactions are reported in Refs. [5-7]. They may be considered as mesoscopic systems. An interesting question is the transition from atomic quantum states to a quasiclassical wavepacket of electrons in bound states. Recently, experiments have been performed to investigate this basic question in atomic physics [8].

A more applied aspect is the possibility to construct a quantum computer basing on the coherent superposition of quantum states, for which also the dephasing time is an important limiting quantity [9].

Our interest is also to find the connection to the standard description of mesoscopic system, such as semiconductor structures, which has been elaborated a few years ago, see, e.g. Imry [10]. In this field of research, some new and interesting experimental results are obtained recently [11]. In particular, we will check the applicability of the Imry's approach [10,12] for the dephasing time given there on the special case of Rydberg atoms interacting with their surroundings. An important question we are interested in is the separation of the system from the environment which is treated as a bath, so back-reactions on the system are neglected.

We give some results for the transition rates of Rydberg atoms. Instead of considering pure quantum states, due to the interaction with the radiation field being always present, transitions to other states occur which result in a finite lifetime of the energy eigenstates. This effect is well known as natural line broadening, it generally demands the quantum description of the radiation field.

If the generation of charged particles in a gas of Ryberg atoms can be suppressed, the strongest interaction is the resonant dipole-dipole multi-particle interaction resulting from the huge $\sim n^2$ dipole matrixelement of the Rydberg atoms. The influence of this $1/R^3$ (R being the distance between two atoms) interaction on the dephasing of a two-atom superposition state has recently been studied [13,14].

Even more effective in destroying the phase of a quantum state are Coulomb collisions. Thus, in Rydberg plasmas where Rydberg atoms are immersed into a plasma consisting of electrons and ions with density n_e, n_i, respectively, the transition rates coincide with the results for dephasing of electron wave packets in mesoscopic physics [10] as long as the interaction is weak. For strong collisions, this approach breaks down, and the dephasing process has to be considered differently.

2. Rydberg Atoms in a Plasma

Starting from a quantum statistical approach, bound states are introduced considering the electron-ion two-particle propagator. The ladder approximation which neglects the influence of the surrounding medium leads to the two-particle Schrödinger equation with a hydrogen-like spectrum of excited bound states. Properties of the plasma such as the equation of state or the transport properties are influenced by the formation of bound states.

Within a virial expansion, the second virial coefficient can be related to the solution of the two-particle problem, i.e. a contribution due to bound states as well as the scattering phase shifts. For short-range potentials, the Beth-Uhlenbeck formula gives the result for the density. The second virial coefficient $n_2 = n_2^{\text{bound}} + n_2^{\text{scatt}}$ consists of a bound state contribution

$$n_2^{\text{bound}}(\beta, \mu) = \sum_{n\mathbf{P}} e^{-\beta(E_{n\mathbf{P}} - \mu_e - \mu_i)} , \qquad (1)$$

where $E_{n\mathbf{P}}$ denotes the energy of a bound state with total momentum $\hbar\mathbf{P}$ and internal quantum number n, and a contribution n_2^{scatt} due to scattering states containing phase shifts that we will not consider further here, see Ref. [15].

For a Coulomb system, the well-known solution of the hydrogen atom leads to an infinite number of bound states, and n_2^{bound} is diverging. This divergence is caused by the large number of weakly bound states, the so-called Rydberg states. Physically, the density as function of temperature and chemical potential is a well-defined finite quantity. As has been shown, e.g. using the Levinson theorem, the divergence is compensated by the contribution of the scattering states. On this background, one can formally introduce the Planck-Larkin partition function

$$n_2^{\text{PL}}(\beta, \mu) = e^{\beta(\mu_e + \mu_i)} \sum_{nP} [e^{-\beta E_{nP}} - 1 + \beta E_{nP}] , \qquad (2)$$

which remains finite. It can argued that the additional terms are traced back to the scattering state contributions. Alternatively, one can introduce a statically screened (Debye) potential so that the number of bound states remains finite, but the continuum edge is shifted downwards so that no so-called transparency windows appears [16].

A more consistent approach would go beyond the two-particle approximation taking into account the interaction with the medium. The bound states are broadened and have a finite lifetime. A two-particle spectral density can be introduced where any sharp distinction between bound and scattering states disappears. The bound states are smeared out and are indistinguishable with the resonances in the continuum. Thus, we have to consider dephasing and homogeneous as well as inhomogeneous broadening when we look for highly excited, quasiclassical states. By homogeneous line-broadening we refer to the effect of actual transitions while inhomogeneous broadening results from averaging over different shifts due to an inhomogeneous background.

In principle, we have always a finite lifetime due to radiative transitions, so that the correct quantum treatment of the Maxwell field already should solve the problem of divergence. Spontaneous emission as well as the interaction with thermal radiation, e.g. the blackbody radiation in equilibrium at given temperature, give transition rates which can be evaluated in dipole approximation for wavelengths large compared with the extension of the bound state. Expressions for transition rates of Rydberg atoms are given in [17], see also [18], and will not be detailed here. In this paper, we will focus to collisions in a plasma since Coulomb interactions are more efficient for broadening under normal plasma conditions, leading to transition rates large compared with the radiative transition rates.

Under certain conditions, other interactions such as the resonant dipole coupling among Rydberg atoms may become of importance, e.g. in Rydberg gases. Hopping processes for

Table 1. Some properties of highly excited (Rydberg) atomic states,
$A_0 = 7.9 \times 10^9 \text{s}^{-1}$ **[17].**

		$n = 10$	$n = 20$	$n = 100$
Ioniz. energy	Ry/n^2	$0.136\,\text{eV}$	$0.034\,\text{eV}$	$1.36\,\text{meV}$
Radius	$a_0 n^2$	$5.3\,\text{nm}$	$21.2\,\text{nm}$	$0.53\,\mu\text{m}$
Trans. freq. $\omega_{n,n\pm1}$	$2\text{Ry}/(\hbar n^3)$	$4.1 \times 10^{13}\text{s}^{-1}$	$5.01 \times 10^{12}\text{s}^{-1}$	$4.1 \times 10^{10}\text{s}^{-1}$
l av. nat. lifetime	$n^5(3\ln n - 0.25)/A_0$	$84\,\mu\text{s}$	$3.54\,\text{ms}$	$17\,\text{s}$

the excitation energy have been investigated recently [19]. All these processes are described within a Coulomb plasma and need for an appropriate treatment of the polarization function. Systematic approaches can be given by diagram representations of the perturbation expansion.

To present a flavor of scales of interest, Tab. 1 gives characteristic properties of Rydberg atoms, see also [20]. As an amazing feature, highly excited Rydberg atoms have dimensions which can be compared with mesoscopic systems, being at the edge to classical systems. Special experimental devices are needed to conserve these very weakly bound systems. Thus, Rydberg (Rb, Xe) atoms with principal quantum number $n \sim 20 \ldots 100$ are generated of laser-cooled atoms in magneto-optical traps with atomic densities $n_A \sim 10^9 - 10^{12}$ cm^{-3} [21-23]. Theoretical description can be found in [24-27]. In absence of external electric fields, seed ionization leads to the generation of ultra-could plasma with $n_e \sim n_i \sim 10^8 - 10^{10}$ cm^{-3}, at temperatures $T_i \sim 1\,\text{K}$, $T_e \sim 20\,\text{K}$. In such plasmas, besides the radiative transitions the most important transition processes involving Rydberg atoms are the scattering of neutral atoms, mostly dipole-dipole interactions among Rydberg atoms and scattering of charged particles.

3. Optical Spectra

The optical properties of a charged particle system are described by the transverse part of the dielectric function. If the wavelength of the light is large compared with the atomic distances, the long-wavelength limit of the dielectric function is related to the refraction index $n(\omega)$ and the absorption index $\alpha(\omega)$ according to

$$\lim_{q \to 0} \varepsilon^{\text{tr}}(\mathbf{q}, \omega) = \lim_{q \to 0} \varepsilon^{\text{long}}(\mathbf{q}, \omega) = \left[n(\omega) + \frac{ic}{2\omega}\alpha(\omega) \right]^2 . \tag{3}$$

In this limit $k \to 0$, the transverse part and the longitudinal part of the dielectric tensor become identical. In particular, we can introduce the longitudinal polarization function

$$\varepsilon^{\mathrm{long}}(\mathbf{q},\omega) = 1 - \frac{1}{\epsilon_0 q^2}\Pi(\mathbf{q},\omega)\,, \tag{4}$$

which is related to the response function $\chi(\mathbf{q},\omega)$ given by

$$[\varepsilon^{\mathrm{long}}(\mathbf{q},\omega)]^{-1} = 1 + \frac{1}{\epsilon_0 q^2}\chi(\mathbf{q},\omega)\,. \tag{5}$$

The response function, in turn, can be expressed in terms of the charge density auto-correlation function or the auto-correlation function of the longitudinal current density,

$$\chi(\mathbf{q},\omega) = \frac{\mathrm{i}\Omega_0}{k_{\mathrm{B}}T}\omega\langle\rho_{\mathbf{q}};\rho_{\mathbf{q}}\rangle_{\omega+\mathrm{i}\eta} = \frac{\mathrm{i}\Omega_0}{k_{\mathrm{B}}T}\frac{q^2}{\omega}\langle\mathbf{j}_{\mathbf{q}};\mathbf{j}_{\mathbf{q}}\rangle_{\omega+\mathrm{i}\eta}\,, \tag{6}$$

with the charge current density $\mathbf{j}_{\mathbf{q}} = \Omega_0^{-1}\sum_{\mathbf{p}c}(e_c/m_c)\hbar q a_{\mathbf{p}+\mathbf{q}}^{\dagger}a_{\mathbf{p}}$ and the current-current correlation function is

$$\langle\mathbf{j}_{\mathbf{q}};\mathbf{j}_{\mathbf{q}}\rangle_{\omega+\mathrm{i}\eta} = \int_{-\infty}^{0} dt\, \mathrm{e}^{\mathrm{i}(\omega+\mathrm{i}\eta)t}\frac{1}{\beta}\int_{0}^{\beta} d\tau\, \mathrm{Tr}\{\rho_0\,\mathbf{j}_{\mathbf{q}}(t-\mathrm{i}\hbar\tau)\mathbf{j}_{-\mathbf{q}}\}\,. \tag{7}$$

The trace is taken with the equilibrium statistical operator ρ_0. The relation to further quantities such as the dynamical structure factor, the dynamical collision frequency or the conductivity will not be given here. Within generalized linear response theory, it can be shown that the use of correlation functions, such as dipole-dipole, is equivalent to the quantum kinetic approach near thermal equilibrium treating the influence of the perturbing plasma on optical transitions in atoms.

Many-particle theory can be applied to evaluate correlation functions in order to derive optical properties. Here, we have to include the formation of bound states, see [1]. In particular, the formation of bound states is taken into account within a cluster decomposition

$$\Pi(\mathbf{q},\omega) = \Pi_1(\mathbf{q},\omega) + \Pi_2(\mathbf{q},\omega) + \dots\,, \tag{8}$$

where $\Pi_1(\mathbf{q},\omega)$ is based on a single-particle loop (RPA) to be improved by self-energy and vertex corrections, and $\Pi_2(\mathbf{q},\omega)$ is based on an atomic loop formed by electron-ion ladder diagrams,

$$\Pi_2^{(0)}(\mathbf{q},\omega) = \sum_{nn'\mathbf{P}} M_{nn'}^{(0)}(\mathbf{q})\frac{g_A(E_{n\mathbf{P}}) - g_A(E_{n'\mathbf{P}+\mathbf{q}})}{E_{n\mathbf{P}} - E_{n'\mathbf{P}+\mathbf{q}} - \hbar\omega + \mathrm{i}0}\,, \tag{9}$$

where $g_A(E_{n\mathbf{P}}) = [\exp((E_{n\mathbf{P}} - \mu_e - \mu_i)/k_{\mathrm{B}}T) - 1]^{-1}$ is the Bose distribution function which gives $n_A\Lambda_A^3/4\exp(-E_{n\mathbf{P}}/k_{\mathrm{B}}T)$ in the non-degenerate limit, using the thermodynamical wavelength $\Lambda_A = [2\pi\hbar^2/(m_A k_{\mathrm{B}}T)]^{1/2}$. Here n_A denotes the density of bound states (atoms) and $m_A = m_e + m_i$ is total mass. For the vertex,

$$M_{nn'}^{(0)}(\mathbf{q}) = -ie\int\frac{d^3\mathbf{p}}{(2\pi)^3}\psi_n^*(\mathbf{p})[\psi_{n'}(\mathbf{p} + \frac{m_e}{m_A}\mathbf{q}) - \psi_{n'}(\mathbf{p} - \frac{m_i}{m_A}\mathbf{q})]$$

$$\approx -ie[\delta_{nn'} - \int\frac{d^3\mathbf{r}}{(2\pi)^3}\psi_n^*(\mathbf{r})e^{\mathrm{i}\mathbf{q}\cdot\mathbf{r}}\psi_{n'}(\mathbf{r})]\,, \tag{10}$$

with $\psi_n(\mathbf{r})$ denoting the eigenfunction in relative coordinates. Depending on n, n', we have bound-bound, bound-free and free-bound as well as free-free contributions, describing optical line spectra, ionization and recombination, as well as bremsstrahlung, respectively. We will focus here only to the discrete part related to the excitation of bound states.

The absorption coefficient is given by the imaginary part of $\Pi(\mathbf{q}, \omega)$ obtained at vanishing denominator, where according to the Dirac identity an imaginary contribution arises. Considering the bound-bound contributions to $\Pi_2(\mathbf{q}, \omega)$ it originates from the transition frequencies between bound states including the Doppler broadening. The correct low-density limit is obtained, where the Coulomb system consists of free charged particles and bound states, and the composition is determined by the Saha equation for chemical equilibrium. Further interactions lead to a modification of the optical spectra. The most efficient process is the interaction with free charged particles, which is the problem we are interested in.

The improvement of the polarization function for free bound states can be given in form of self-energies and vertex corrections. The two-particle (atomic) propagator is given by the spectral function as

$$G_2(n, P, \Omega_\lambda) = \int \frac{d\omega}{2\pi} \frac{A_2(n, P, \omega)}{\Omega_\lambda - \omega} \tag{11}$$

and

$$A_2(n, P, \omega) = -2i \frac{\mathrm{Im}\Sigma_2(n, P, \omega)}{(\hbar\omega - E_{nP} - \mathrm{Re}\Sigma_2(n, P, \omega))^2 + (\mathrm{Im}\Sigma_2(n, P, \omega))^2}. \tag{12}$$

In general, the self-energy $\Sigma_2(n, P, \omega)$ is not diagonal in the internal quantum states n so that matrix relations have to be applied.

Further diagrams describing interactions between the two atomic propagators are condensed into a vertex correction. Systematic approximation schemes can be given such that self-energy and vertex corrections are treated on the same level, as well known from conserving approximations obeying the Ward-Takahashi relation.

A systematic treatment of the influence of a dense plasma on the shape of spectral lines can be given by theories according to Griem [28], the unified theory, etc. We use a systematic approach based on the treatment of the dielectric function. The final expressions are (see Günter [29] for details)

$$\mathcal{L}(\omega) \sim \sum_{nn'mm'} I_{nm}^{n'm'}(\omega) \int \frac{d^3\mathbf{P}}{(2\pi)^3} \exp(-\frac{\hbar^2\mathbf{P}^2}{2k_\mathrm{B}Tm_A}) \int_0^\infty d\bar{\beta} \, W_\rho(\bar{\beta})$$

$$\times \mathrm{Im}\langle n|\langle m| \left[\hbar(\omega - \omega_{nm}) - \frac{\hbar^2\mathbf{P}\cdot\mathbf{q}}{m_A} - \frac{\hbar^2\mathbf{q}^2}{2m_A} - \mathrm{Re}\left(\Sigma_n(\omega, \bar{\beta}) - \Sigma_m(\omega, \bar{\beta})\right) \right.$$

$$\left. - i\,\mathrm{Im}\left(\Sigma_n(\omega, \bar{\beta}) + \Sigma_m(\omega, \bar{\beta})\right) + i\Gamma_{nm}^\mathrm{V} \right]^{-1} |m'\rangle|n'\rangle, \tag{13}$$

where $I_{nm}^{n'm'}(\omega) = \omega^4/(8\pi^3 c^3) \exp(-\omega/k_\mathrm{B}T) M_{nm}^{(0)}(\mathbf{q}) [M_{n'm'}^{(0)}(\mathbf{q})]^*$ are the frequency dependent intensities of emission and $\omega_{nm} = (E_n - E_m)/\hbar$. $W_\rho(\bar{\beta})$ is the distribution function of the ionic microfield with $\bar{\beta} = |\mathbf{E}|/E_\mathrm{H}$, introducing the Holtsmark field-strength

$E_{\mathrm{H}} = e/(4\pi\varepsilon_0 r_0^2)$, r_0 given by $(4/15)(\sqrt{2\pi}r_0)^3 n_i = 1$. The coupling term Γ_{nm}^V of the upper and lower level, $|n\rangle$ and $|m\rangle$, respectively plays an important role for transition among highly excited states [30], but not if initial and final levels are well separated what we will consider in the following. In Eq. (13) the ionic contribution goes into the microfield while the effect of the electron is considered by impact collision. The shift of the upper levels is the important one and is given by the electronic self-energy contribution

$$\Delta_n^{\mathrm{SE}} + i\Gamma_n^{\mathrm{SE}} = < n|\Sigma^{\mathrm{el}}(E_n^0/\hbar + \Delta\omega, \bar{\beta})|n> = -\int \frac{d^3\mathbf{q}}{(2\pi)^3} V(q) \sum_\alpha |M_{n\alpha}^{(0)}(\mathbf{q})|^2$$

$$\times \int_{-\infty}^{\infty} \frac{d\hbar\omega}{\pi}(1 + n_B(\omega))\frac{\mathrm{Im}\,\varepsilon^{-1}(\mathbf{q}, \omega + \mathrm{i}0)}{\hbar\Delta\omega + E_n^0 - E_\alpha(\bar{\beta}) - (\hbar\omega + \mathrm{i}0)} \quad, \tag{14}$$

where $V(q) = e^2/(\varepsilon_0 q^2)$ is the Fourier transform of the Coulomb potential and $n_B(\omega) = [\exp(\hbar\omega/k_{\mathrm{B}}T) - 1]^{-1}$ denotes the Bose function. The frequency ω of Eq. (13) is taken near the free atom's eigenvalue E_n^0, i.e. small $\Delta\omega = \omega - E_n^0/\hbar$. Contributions to the line-shifts are not obtained in no-quenching approximation, but are taken from transitions between different principal quantum numbers $n \neq \alpha$. Furthermore, the following approximations for the dielectric functions are taken,

$$\varepsilon(\mathbf{q}, \omega) \approx \varepsilon^{\mathrm{RPA}}(\mathbf{q}, \omega) = 1 - 2V(q)\int \frac{d^3\mathbf{p}}{(2\pi)^3}\frac{f_e(E_{\mathbf{p}}) - f_e(E_{\mathbf{p}+\mathbf{q}})}{E_{\mathbf{p}} - E_{\mathbf{p}+\mathbf{q}} - (\hbar\omega + \mathrm{i}0)} \quad, \tag{15}$$

$$\mathrm{Im}\,\varepsilon^{-1}(\mathbf{q}, \omega) \approx -\mathrm{Im}\,\varepsilon(\mathbf{q}, \omega) = 2\pi V(q)\int \frac{d^3\mathbf{p}}{(2\pi)^3}[f_e(E_{\mathbf{p}}) - f_e(E_{\mathbf{p}+\mathbf{q}})]\delta(E_{\mathbf{p}} - E_{\mathbf{p}+\mathbf{q}} - \hbar\omega), \tag{16}$$

$$[1 + n_B((E_{\mathbf{p}} - E_{\mathbf{p}+\mathbf{q}})/\hbar)][f_e(E_{\mathbf{p}}) - f_e(E_{\mathbf{p}+\mathbf{q}})] = -f_e(E_{\mathbf{p}+\mathbf{q}})[1 - f_e(E_{\mathbf{p}})] \approx -f_e(E_{\mathbf{p}+\mathbf{q}}), \tag{17}$$

using the shorthand notation for the free particle's energy $E_{\mathbf{p}} = \hbar^2 p^2/(2m_e)$. All together, the electronic self-energy results in [29]

$$\Delta_n^{\mathrm{SE}} = -2\sum_\alpha \mathcal{P}\int \frac{d^3\mathbf{q}}{(2\pi)^3}\frac{d^3\mathbf{p}}{(2\pi)^3}V^2(q)f_e(E_{\mathbf{p}})\frac{|M_{n\alpha}^0(\mathbf{q})|^2}{\hbar\omega_{n\alpha} + \frac{\hbar^2\mathbf{p}\cdot\mathbf{q}}{m_e} + \frac{\hbar^2 q^2}{2m_e}} \quad,$$

$$\Gamma_n^{\mathrm{SE}} = 2\pi\sum_\alpha \int \frac{d^3\mathbf{q}}{(2\pi)^3}\frac{d^3\mathbf{p}}{(2\pi)^3}V^2(q)f_e(E_{\mathbf{p}})|M_{n\alpha}^0(\mathbf{q})|^2\delta\left(\hbar\omega_{n\alpha} + \frac{\hbar^2\mathbf{p}\cdot\mathbf{q}}{m_e} + \frac{\hbar^2 q^2}{2m_e}\right), \tag{18}$$

with $\hbar\omega_{n\alpha} = E_\alpha(\bar{\beta}) - E_n^0 - \hbar\Delta\omega$. Considering the non-degenerate case, the integration over the electron momentum leads to the following expressions,

$$\Delta_n = -\frac{n_e e^4}{2\pi^3 \hbar\varepsilon_0^2}\sqrt{\frac{m_e}{2k_{\mathrm{B}}T}}\sum_\alpha \int \frac{d^3\mathbf{q}}{q^5}|M_{n\alpha}^0(\mathbf{q})|^2 F(x) \quad,$$

$$\Gamma_n = \frac{n_e e^4}{8\pi^2 \hbar\varepsilon_0^2}\sqrt{\frac{m_e}{2\pi k_{\mathrm{B}}T}}\sum_\alpha \int \frac{d^3\mathbf{q}}{q^5}|M_{n\alpha}^0(\mathbf{q})|^2 \exp(-x^2) \quad, \tag{19}$$

where

$$F(x) = \mathrm{e}^{-x^2} \int_0^x \mathrm{e}^{t^2}\,dt\,, \qquad x = \frac{1}{2\hbar}\sqrt{\frac{2m_e}{k_{\mathrm{B}}T}}\left(\frac{\hbar\omega_{n\alpha}}{q} + \frac{\hbar^2 q}{2m_e}\right)\,, \tag{20}$$

is the Dawson integral. For $\Delta\omega = 0$ and $E_\alpha(\bar{\beta}) = E_\alpha^0$ we have the impact approximation.

4. Mean Field and Ionic Microfield

The influence of the medium on the quantum system, i.e. the Rydberg atom, can be treated within perturbation theory. In lowest order, the influence of the medium is given by the Hartree-Fock approximation which neglects any correlations in the medium. The Hartree-Fock approximation is instantaneous, without any retardations which would produce a frequency dependence of the dynamic interaction. However, instantaneous contributions are also obtained from correlated distributions, and the cluster-Hartree Fock approximation [31] has been introduced to determine the optimum Hamiltonian. Within a Dyson equation approach, the instantaneous part of the effective Hamiltonian was found according to [32]

$$H_{\alpha\beta}^{(0)} = \sum_\gamma \langle[[A_\alpha, H], A_\gamma^\dagger]_\pm\rangle\{\langle[A_\gamma, A_\beta^\dagger]_\pm\rangle\}^{-1}\,, \tag{21}$$

where $A_\alpha^\dagger$, A_α are creation and annihilation operators for clusters.

As an important point, no dephasing arises in the Hatree-Fock approximation. However the solution of the Schrödinger equation for an arbitrary potential, like the impurity problem, may lead to an inhomogeneous broadening of energy levels, which is not related to any finite lifetime. With respect to the Rydberg plasma considered here, in adiabatic approximation the ions are considered as fixed at random positions, given by a pair distribution function. A nearly static microfield arises, producing a fluctuating Stark shift of the energy levels. The corresponding Stark broadening, already included in Eq. (13) as an average over the microfield distribution $W_\rho(\bar{\beta})$, is not connected with a dephasing process. The solution of the Schrödinger equation leads to sharp energy levels, distributed on energy in correspondence to the microfield distribution.

Starting from free particle states, the Hartree term is divergent but disappears because of charge neutrality. On the other hand, in the limit of large masses, i.e. in the adiabatic limit, we can pass to a quasiclassical description where we introduce localized states. The introduction of (relatively stable) wave packets is possible for ions, but not relevant for electrons where the wave function spreads out during the collision or transition time.

In general, the transition to a classical behavior is of interest also for the weakly bound electron in the Rydberg atom. Superposing different Rydberg eigenstates, stable wave packets have been considered [33-35]. They are prepared in laser excited Rydberg systems [8]. It is argued [2] that localized states become relevant for the coupling of the system to a bath, when the coupling contains the position operator. In this representation with respect to localized states, the density matrix becomes nearly diagonal. Dephasing destroys the non-diagonal elements of the density matrix and, in the case of collisional interaction, will diagonalize the density matrix. The localized wave-packets are the objects for which the transition to quasiclassical description can be performed.

5. Electronic Scattering

We now consider an atom in a highly excited Rydberg state in a cold, dilute electron-ion plasma that has been studied in several experiments [21]. We choose the typical setup where we have an electronic part of the plasma with temperature $T = 20$ K and density $n_e = 10^9$ cm^{-3}. A further characterizing property of such a plasma is the plasma frequency, here $\omega_{\rm pl} = \sqrt{n_e e^2/(\varepsilon_0 m_e)} = 8.6 \times 10^9\,{\rm s}^{-1}$. The Coulomb coupling parameter is $\Gamma = \frac{e^2}{4\pi\varepsilon_0 k_{\rm B}T}(4\pi n_e/3)^{1/3} = 0.135$, and the degeneracy parameter $\Theta = 2m_e k_{\rm B}T/\hbar^2\,(3\pi^2 n_e)^{-2/3} = 5 \times 10^5$. Such a plasma can clearly be described as non-degenerate.

Transitions among principal quantum numbers n, n' are obtained by averaging over l and summing over l'. The matrix elements in Eq. (10) among n, $\alpha = n' \neq n$ can for high n be expressed by their quasiclassical limit (introducing atomic radius a_0) [36],

$$|M_{nn'}^{(0)}(\mathbf{q})|^2 = \frac{512}{3\pi n^3 (n')^3}\,\frac{\chi^5}{(\chi^4 + 2\chi^2(\epsilon + \epsilon') - (\triangle\epsilon)^2)^3}\,,$$

$$\chi = a_0|q|\,,\quad \epsilon = 1/n^2\,,\quad \epsilon' = 1/(n')^2\,,\quad \triangle\epsilon = \epsilon' - \epsilon\,,\tag{22}$$

which perfectly agrees with the quantum mechanical expression for $n > 5$ [37].

These matrix elements, on the other hand, can be used to calculate the total cross-section $\sigma_{nn'}(E_p)$ in first Born approximation (BA) for $n \to n'$ transitions induced by an incident electron of energy E_p, which in turn can be averaged over a non-degenerate electron gas to obtain the transition rates $W_{nn'}$ [38]. The expression for excitation $n' - n = \triangle n > 0$ with transition frequency $\hbar\omega_{nn'} = {\rm Ry}\,(1/(n')^2 - 1/n^2)$ is

$$W_{nn'}(T) = n_e\langle\sqrt{2E_p/m_e}\,\sigma_{nn'}(E_p)\rangle_T\,,$$

$$\sigma_{nn'}(E_p) = \frac{8\pi{\rm Ry}}{E_p}\int_{\sqrt{2m_e/h^2}|\sqrt{E_p}-\sqrt{E_p+\hbar\omega_{nn'}}|}^{\sqrt{2m_e/h^2}(\sqrt{E_p}+\sqrt{E_p+\hbar\omega_{nn'}})}\frac{M_{nn'}^2(q)\,dq}{q^3}\,.$$

$$\tag{23}$$

A corresponding formula holds for deexcitation. $\langle\ldots\rangle_T$ denotes averaging the incident electron energy E_p over a Maxwell-Boltzmann distribution of temperature T. This distribution is not exactly valid for the velocities of electrons in the geometrical setup found in experiments on ultra-cold plasmas which can more accurately described by a Kramers-Michie-King distribution, but the Boltzmann distribution used here is a sufficient approximation not too far off the center of the plasma [39].

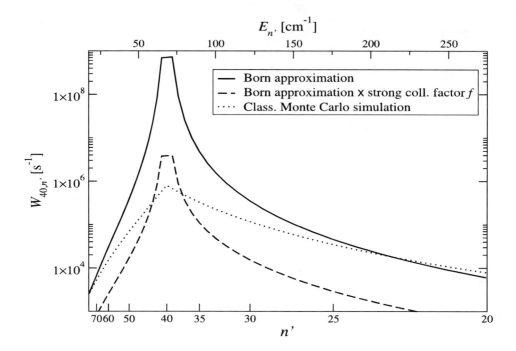

Figure 1. Transition rates of $n = 40$ to near n' states induced by a $T = 20$ K electron plasma with density $n_e = 10^9 \, \mathrm{cm}^{-3}$ for the BA with (solid line) and without (dashed line) effect of strong collisions [17] compared with CTMC calculation [40] (dotted line).

The transition rates obtained for high n in the BA strongly exceed these obtained classical trajectory Monte-Carlo (CTMC) simulations [40] as well as they are damped if one includes the effects of strong collisions [17], see Fig. 1 for $n = 40$. This is already effective at $n = 13$ where it can be underlined by experimental results [5,18].

Always the transitions to adjacent n' dominate though the effect is more pronounced for the quantum calculation independently of inclusion of part of strong collisions. Experiments have shown [5] that agreement of the CTMC rates with the data is generally better compared to quantum calculations. This is not only a consequence of the limited treatment of strong collision in [17]. One has to take into account, that the limiting correspondence in quantum calculations for the case of the CTMC calculations are wave-packets with vanishing width and not the energy eigenstates of the atomic Hamiltonian. The good agreement of the experiments and the CTMC rates indicates that the prepared initial (and final) states as well as the intermediate (virtual) states in the treatment of strong collisions correspond to the classical one.

On the other hand, in the BA there is a direct relation between the total transition

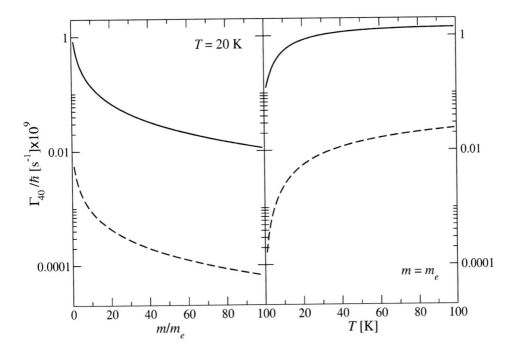

Figure 2. Total rates (BA solid lines, BA with strong collision factor f dashed lines) of $n = 40$ quickly decrease at increasing mass m (left) when $T = 20$ K constant. Same rates increase with increasing temperature T but reach saturation for $m = m_e$ at $T \approx 100$ K.

rate and the imaginary part of the self-energy in the approximation of Eq. (19),

$$\sum_{n'} W_{nn'} = \frac{2}{\hbar} \Gamma_n(\Delta\omega = 0). \tag{24}$$

The line broadening due to electron impact is in the limiting case of weak collision fully related to the lifetime of the upper state. Additionally Fig. 1 shows that the sum of α in Eqs. (14-19) can always be taken as effectively finite.

In Fig. 2 we show the dependence of the transition rates on the mass of the scattering particle as well as on the temperature in the range appropriate for ultra-cold plasmas.

6. Dephasing Rate and Lifetime

Our aim is to combine different approaches to dephasing processes and corresponding lifetimes of quantum states. The scattering process averaged over the distribution of scattering particles of the last section is a special case of a system-background picture known from the studies of open quantum systems [3] where there is the system under consideration

(S) that is coupled by an interaction potential V_I to its environment called background (B) that is not accessible to measurements. The total Hamiltonian of such a system reads

$$H = H_{\mathrm{S}} + H_{\mathrm{B}} + V_{\mathrm{I}} \,. \tag{25}$$

An important effect of the background on the system is the suppression of non-diagonal elements in the reduced density operator of the system (where the degrees of freedom of the environment have been traced out) in a basis where the interaction potential operator is diagonal [2]. This effect that suppresses the detection of superposition states in that basis is known as decoherence.

The process of decoherence can be studied in systems of mesoscopic scales as Rydberg atoms are. In [12] it was shown that in the framework of linear-response theory (LRT) any transition due to the system-bath interaction randomizes the phase of the system if the background has a very high quantum uncertainty. In that case the background is also referred to as heat bath that 'dephases' the system. Then the sum of transitions can be taken as a measure of decoherence. One defines the dephasing time τ_ϕ as a scale on that interferences are suppressed. It can be estimated by the dephasing time formula [10],

$$\frac{1}{\tau_\phi} = \frac{1}{\hbar^2 (2\pi)^2 \Omega_0} \int d\mathbf{q} \int\limits_{-\infty}^{\infty} d\omega \, |V_q|^2 S_{\mathrm{S}}(-\mathbf{q}, -\omega) S_{\mathrm{B}}(\mathbf{q}, \omega) \,. \tag{26}$$

Both, the system and the background enter Eq. (26) by their dynamical structure factor (DSF). This is a well known quantity in LRT containing the charged particle density $\rho_{\mathbf{q}}(\mathbf{r})$, the many-particle states $|\phi_e\rangle$, and their statistical weight P_{ϕ_e}. For $S_{\mathrm{B}}(\mathbf{q}, \omega) = S_e(\mathbf{q}, \omega)$ we have

$$S_e(\mathbf{q}, \omega) = \sum_{\phi_e, \phi_e'} |\langle \phi_e' | \rho_{\mathbf{q}} | \phi_e \rangle|^2 \delta(\omega + \omega_{\phi_e \phi_e'}) P_{\phi_e} \,. \tag{27}$$

From the fluctuation dissipation relation [1],

$$S_e(\mathbf{q}, \omega) = \frac{\hbar \Omega_0}{\pi} \frac{1}{e^{\beta \hbar \omega} - 1} \frac{\varepsilon_0 q^2}{e^2} \, \mathrm{Im} \left\{ \varepsilon_e^{-1}(\mathbf{q}, \omega) \right\} \,, \tag{28}$$

the DSF can be calculated from the dielectric function.

On the other hand, if we assume the system to be a single (but for the background) isolated Rydberg atom, the DSF of the system can be expressed by the atomic matrix element,

$$S_{\mathrm{S}}(-\mathbf{q}, -\omega) = \sum_{\alpha \alpha'} |M_{\alpha \alpha'}^{(0)}(-\mathbf{q})|^2 \delta(\omega_{\alpha \alpha'} - \omega) P_\alpha \,, \tag{29}$$

where the sum runs over the considered atomic eigenstates. Then we obtain for the dephasing time

$$\frac{1}{\tau_\phi} = \sum_{\alpha \alpha'} P_\alpha \frac{e^2}{(4\pi\varepsilon_0)\hbar\pi^2} \int \frac{d^2\mathbf{q}}{q^2} |M_{nn'}^{(0)}(q)|^2 \frac{1}{e^{\beta\hbar\omega_{\alpha\alpha'}} - 1} \mathrm{Im} \left\{ \varepsilon_e^{-1}(\mathbf{q}, \omega_{\alpha\alpha'}) \right\} \,. \tag{30}$$

If we use the dielectric function in RPA using Eq. (16) for the non-degenerate case and consider a distribution $P_\alpha = \delta_{\alpha n}$ for the initial state of the Rydberg atom, we arrive again at the total transition rate in the BA. Note that we arrive at Eq. (26) as well when we calculate the lifetime of a Rydberg atom in a many-particle environment using the total state of the form $|\Psi\rangle = |nlm\rangle \otimes |\phi_e\rangle$ and Fermi's Golden Rule [18]. An obvious advantage of this approach is, that additional many-particle effects originating from inter-electron collisions can be included into the rates [41]. In this case one obtains the rates for weak scattering of a non-ideal electron gas at a Rydberg atom.

This result shows that Eq. (26) can be understood as weakly coupling of the system and the background, that is appropriate as long as the BA gives the correct transition rates. Altogether, different approaches describing the transition rates and lifetime of Rydberg states due to electron interaction in the weak scattering limit give the same result. In particular, cross sections in BA are compared here with the impact approximation for the line shape of optical transitions and an expression for the lifetime of excited states in mesoscopic systems.

The BA is restricted to weak scattering processes. In the case of strong scattering, we have to modify the results obtained for transition rates and lifetimes. As shown in the following section, corresponding improvements are known for the electron-atom cross section. Also, strong collisions are considered in the unified theory of line profiles. However, the treatment of strong coupling between system and background in the dephasing time formula, Eq. (26) is rather unclear as we point out in the following section. To the knowledge of the authors no improvements towards higher orders in the perturbation formalism for dephasing in the last section has yet been made.

7. Strong Collisions

We start again with the cross section describing the collision between a charged particle (electron) and the atom. Up to now, we have used the solution of the two-particle problem such as the Rydberg state of the atom and considered the collision with the plasma electrons in the BA. Considering the case that only one particle is strongly interacting with the excited atom, we have to solve the three-particle problem embedded in the surrounding plasma. In principle, we have to consider a three-particle T-matrix approximation which leads to the solution of an in-medium Faddeev equation.

Different methods and approximations have been worked out to find the solution of the three-particle problem such as the close-coupling method [42]. As an approximative solution of this problem, the impulse approximation, see [43], has been worked out for the case when the assumption of weak interaction between perturber and bound electron is no longer valid but the binding interaction is negligible. Here, the bound electron interacts with the charged perturber as described by a two-particle T-matrix. For highly excited states, these higher order effects can be calculated in a classical correspondence framework [17], where a fit-formula is given, that contains the damping of the Born cross section by higher order effects,

$$\sigma_{nn'}(E_p) = \sigma_{nn'}(E_p)^{(\mathrm{BA})} \times f(n, \triangle n; \theta),$$

$$f(n, \triangle n; \theta) = \frac{\ln[1 + \frac{1}{\triangle n\, \theta(1+2.5n/(\triangle n\theta))}]}{\ln[1 + \frac{1}{\triangle n\, \theta}]}, \quad \theta = \sqrt{\frac{|E_n|}{k_{\mathrm{B}} T}}. \qquad (31)$$

The transition rates obtained in BA are significantly reduced if strong collisions are taken into account. As shown from Fig. 1 for the transition rates from the initial state $n = 40$, the account of strong collisions leads to a substantial modification of the Born result. This result coincides with results considering transition rates from $n = 13$ [18], where the reduction of transition rates in BA leads to good agreement with experimental results. Moreover for both initial values of n the agreement with the total rates calculated by CTMC simulation is very good, though the quantum rates are higher for small Δn and lower for greater Δn. As pointed out in section 5, the quantum correspondence of the CTMC calculations are non-dispersing wave-packet dynamics.

The same reasoning also applies to the Eq. (26) which corresponds to the BA. Therefore the results for the dephasing time becomes invalid if strong collisions are of relevance. If we deduce, e.g., the decoherence time according to the mesoscopic procedure ignoring strong collision we arrive at a dephasing time around 1 ns for a $n = 40$ level. Taking into account the reduction of the transition rates by the already discussed effect of strong collisions would even increase the dephasing time by about two orders of magnitude. However, strong collision have to be considered in the representation of the reduced system. Therefore a different treatment of the effect of strong and weak perturbers is necessary.

The problem that the BA overestimates the contribution of the perturbers is also known calculating the effect of electron impact to the spectral line profile. The width according to the BA is reduced introducing a semi-empirical cut-off procedure [28]. This procedure is motivated by the distinction between weak collisions at collision parameter values large compared with the Weisskopf radius R_W,

$$R_W = \left(\frac{\alpha_p C_p}{V}\right)^{1/(p-1)}, \quad \alpha_p = \sqrt{\pi} \frac{\Gamma[(p-1)/2]}{\Gamma(p/2)}, \qquad (32)$$

and strong collisions at small collision parameters. The Weisskopf radius is here defined by the impact parameter that shifts the phase of a classical colliding particle by unity, and $p = 2$, $\alpha_2 = \pi$ for the case of Coulomb scattering and linear Stark effect (hydrogen). Though the line-constant C_2 was originally derived in a classical theory, it can be calculated for the quantum case giving [30]

$$C_2 = \frac{\hbar}{m_e}\sqrt{\frac{2}{3}I(n, n')}, \quad I(n, n') = I(n) + I(n') + K(n, n'). \qquad (33)$$

The line intensities $I(n, n')$ are the normalized dipole matrixelements of the transitions associated with the considered line. For our case of a line between a highly excited Rydberg state with n and a state near the ground state, $n' \sim 1$, $I(n)$ gives the strongest part, and $I(n) \sim n^4$. Remarkably the Weisskopf radius scales with the atomic radius $n^2 a_0$ of the Rydberg atom. For an electron gas at temperature $T = 20\,\mathrm{K}$, the average velocity is $V \sim 0.01$ a.u., resulting in a Weisskopf radius of about a hundred times the radius of atom in the upper state. If we consider a density of $n_e = 10^9\,\mathrm{cm}^{-3}$ the mean distance of the electrons equals the Weisskopf radius for a transition of $n \sim 30$ to a state near the ground state, showing that for this case the collisions are partially strong.

Analogous to the damping factor Eq. (31), Griem [28] introduces a cut-off parameter which selects out the strong collision contributions from the BA. Following this procedure, the homogeneous impact broadening of optical lines is reduced leading to a longer dephasing time compared with the result of Eq. (26) as already discussed above. This simple procedure cannot generally be accepted for the derivation of Eq. (26) for the dephasing time, because strong collisions have a massive effect on the phases. It should be mentioned here that, nevertheless, within LRT the semi-empirical introduction of a cut-off procedure can be replaced by the systematic treatment of T matrices [44].

For the ionic case the larger mass leads to a factor of $R_{W,i} = 43\, R_{W,e}$ for the case of protons in a hydrogen plasma and even higher for larger atoms assuming thermal equilibrium between electrons and ions. In experiments on ultra-cold plasmas the average velocity (temperature) of the ions is lower than the velocity of the electrons and that factor by its square root again increases the Weisskopf radius for ionic collisions. That means, a large fraction of the collision processes are strong and, because of multiple ions being within the Weisskopf radius of the Rydberg atom, a single-scatterer picture can no longer be valid. On the other hand, Fig. 2 shows that the transition rates are strongly reduced for larger masses.

For the linewidth calculation the special character of the ionic contribution has been taken into account in Eq. (13) by the microfield concept. It also gives a reduced broadening of the line shape, but as well it modifies the atomic states used in the calculation of the electron impact approximation rates. In principle, we have an inhomogeneous broadening due to the fluctuating microfield instead of the homogeneous broadening of the electron impact approximation. Only the homogeneous broadening is related to the life time or to the dephasing time of the excited state.

As a consequence, Eq. (26) cannot be valid in the case of strong collisions. Instead of this, the separation of the system and the bath has to be reconsidered in a way that strongly interacting particles together with the atom form the system, or, in other words, only dephasing of atomic states that include the influence of strong scatterers can be considered by the approach.

This is well known from mesoscopic physics, where wave packets of electrons that diffuse between fixed perturbing ions (impurities), which do not contribute to dephasing, are considered. Only the coupling to dynamic degrees of freedom, which can absorb energy, leads to decoherence [10]. Therefore, the limit of large masses is also treated by strong collisions. In the adiabatic limit, the dephasing time is increasing.

8. Concluding Remarks

Transitions to near levels by collision with charged particles are the dominating processes for Rydberg atoms in an ultra-cold dilute plasma. Weak collisions can be treated in Born approximation, the resulting transition rates give lifetimes in agreement with the expressions obtained for the dephasing rate in mesoscopic systems. On the other hand, the account of strong collisions leads to a reduction of the transition rates, and the simple expression for dephasing time in mesoscopic systems is not longer valid.

A similar behavior is known from the theory of pressure broadening of spectral lines. The impact approximation can be related to the Born approximation in the case of weak scattering. Strong collisions are of relevance when the mean distance among the charged particles and the atom is of the order of the Weisskopf radius, which scales with the size of the Rydberg atom for transitions to low energy states. They become dominant if the mass of the perturbing charge is sufficiently large so that the concept of a static microfield is applicable. Thus, the Born approximation is no longer sufficient and has to be improved considering T-matrix approaches. Once more, this is not described by the lifetime formula, Eq. (26), for mesoscopic systems. In general, the decomposition of the system interacting with a bath has to be reconsidered including all strongly interacting particles into the system. Then, the coupling with the remaining bath is weak so that no back-reaction originates from the bath, and the simple expression for dephasing time in mesoscopic systems is expected to be valid with respect to this new subdivision.

Another interesting topic is the transition to classical behavior. For this, coherent superpositions of energy eigenstates should be used to form wave packets which are localized. The dephasing properties of such localized states will be the subject of future work. In general, the optimal basis to represent the density matrix of the system is an important problem to describe the dephasing of non-diagonal elements.

Acknowledgements

One author (CG) gratefully acknowledges the support by SFB 652 of the DFG.

References

[1] W.-D. Kraeft, D. Kremp, W. Ebeling, and G. Röpke, *Quantum Statistics of Charged*

Particle System (Plenum, New York, 1986) [or Akademie Verlag, Berlin 1986].

[2] E. Joos, H.D. Zeh, C. Kiefer, D. Giulini, J. Kupsch, and I.-O. Stamatescu, *Decoherence and the Apperrance of a Classical World in Quantum Theory*, 2nd ed. (Springer, Berlin, 2003).

[3] U. Weiss, *Quantum Dissipative Systems*, 2nd ed. (World Scientific, Singapore, 1999).

[4] G. Auletta, *Foundations and Interpretation of Quantum Mechanics* (World Scientific Publishing Co. Pte. Ltd., Singapore, 2000).

[5] F. Devos, J. Boulmer, and J.-F. Delpech, *J. Phys. (Paris)* **40**, 215, (1979).

[6] G.W. Foltz, E.J. Beiting, T.H. Jeys, K.A. Smith, F.B. Dunning, and R.F. Stebbings, *Phys. Rev.* **A 25**, 187, (1982).

[7] K.B. MacAdam, D.A. Crosby, and R. Rolfes, *Phys. Rev.* **A 24**, 1286, (1981); K.B. MacAdam, R. Rolfes, X. Sun, J. Singh, W.L. Fuqua III, and D.B. Smith, *Phys. Rev.* **A 36**, 4245, (1987).

[8] M.W. Noel and C.R. Stroud, *Phys. Rev. Lett.* **77**, 1913, (1996); Z.D. Gaeta, M.W. Noel, and C.R. Stroud, *Phys. Rev. Lett.* **73**, 636, (1994); M. Mallalieu and C.R. Stroud, *Phys. Rev.* **A 49**, 2329, (1994); Z.D. Gaeta and C.R. Stroud, *Phys. Rev.* **A 42**, 6308, (1990); J.A. Yeazell and C.R. Stroud, *Phys. Rev.* **A 43**, 5153, (1991).

[9] D.P. Vincenzo, *Science* **270**, 255. J. Gruska, *Quantum Computing* (McGraw-Hill, London, 1999); A.M. Steane, *Rep. Prog. Phys.* **61**, 117, (1998); I.I. Ryabtsev, D.B. Tretnyakov, and I.I. Beterov, *J. PHys.* **B 38**, S421, (2005).

[10] Y. Imry, *Introduction to Mesoscopic Physics*, 2nd ed. (Oxford University Press, Oxford, 2002).

[11] J.E. Mooij, T.P. Orlando, L. Levitov, L. Tian, C.H. van der Wal, and S. Lloyd, *Science* **285**, 1038, (1999); T.P. Orlando, J.E. Mooij, L. Tian, C.H. van der Wal, L.S. Levitov, S. Lloyd, and J.J. Mazo, *Phys. Rev.* **B 60**, 15398, (1999); I. Chiorescu, Y. Nakamura, C.J.P.M. Harmans, and J.E. Mooij, *Science* **299**, 1869, (2003); C.H. van der Wal, F.K. Wilhelm, C.J.P.M. Harmans, and J.E. Mooij, *Eur. Phys. J.* **B 31**, 111, (2003); F.K. Wilhelm, M.J. Storcz, C.H. van der Wal, C.J.P.M. Harmans, and J.E. Mooij, *Adv. Solid State Phys.* **43**, 763, (2003).

[12] A. Stern, Y. Aharonov, and Y. Imry, *Phys. Rev.* **A 41**, 3436, (1990).

[13] W.R. Anderson, M.P. Robinson, J.D.D. Martin, and T.F. Gallagher, *Phys. Rev.* **A 65**, 063404, (2002).

[14] M. Mudrich, N. Zahzam, T. Vogt, D. Comparet, and P. Pillet, *arXiv:physics/0504022*.

[15] M. Schmidt, G. Röpke, and P. Schuck, *Ann. Phys. (N.Y.)* **202**, 57, (1990).

[16] F.E. Höhne, R. and Zimmermann, *J. Phys.* ,**B 15**, 2551, (1982).

[17] I.L. Beigman and V.S. Lebedev, *Phys. Rep.* **250**, 95, (1995); I.L. Beigman and V.S. Lebedev, *Physics of Highly Excited Atoms and Ions* (Springer, Berlin, 1998).

[18] C. Gocke, and G. Röpke, *submitted to J. Phys. A*, SCCS05 Proc. special edition.

[19] F. Robicheaux, J.V. Hernandez, T. Topcu, and L.D. Noordam, *Phys. Rev.* **A 70**, 042703, (2004).

[20] T.F. Gallagher, *Rydberg Atoms* (Cambridge, University Press, 1994).

[21] T.C. Kilian, S. Kulin, S.D. Bergeson, L.A. Oronzco, C. Orzel, and S.L. Rolston, *Phys. Rev. Lett.* **83**, 4776, (1999); S. Kulin, T.C. Kilian, S.D. Bergeson, and S.L. Rolston, *Phys. Rev. Lett.* **85**, 318, (2000); T.C. Kilian, M.J. Lim, S. Kulin, R. Dumke, S.D. Bergeson, and S.L. Rolston, *Phys. Rev. Lett.* **86**, 3759, (2001).

[22] E. Eyler, A. Estrin, J.R. Ensher, C.H. Cheng, C. Sanborn, and P.L. Gould, *Bull. Am. Phys. Soc.* **45** 56, (2000).

[23] M.P. Robinson, B.L. Tolra, M.W. Noel, T.F. Gallagher, and P. Pillet, *Phys. Rev. Lett.* **85**, 4466, (2000).

[24] F. Robicheaux, and J.D. Hanson, *Phys. Rev. Lett.* **88**, 055002, (2002).

[25] S.G. Kuzmin, and T.M. O'Neil, *Phys. Rev. Lett.* **88**, 065003, (2002).

[26] A.N. Tkachev and S.I. Yakovlenko, *Quantum Electronics* **31**, 1084, (2001).

[27] T. Pohl, T. Pattard, and J.M. Rost, *Phys. Rev.* **A 70**, 033416, (2004).

[28] H.R. Griem, *Principles of Plasma Spectroscopy*, (Cambridge Univ. Press, Cambridge, 1997).

[29] S. Günter, *habil. thesis* (Universität Rostock, 1995).

[30] I.I. Sobelman, L.A. Vainshtein, and E.A. Yukov, *Excitation of Atoms and Broadening of Spectral Lines*, (Springer, Berlin, 1981).

[31] J. Dukelsky, G. Röpke, and P. Schuck, *Nucl. Phys.* **A 628**, 17, (1998).

[32] G. Röpke, A. Schnell, and P. Schuck, in *Condensed Matter Theories*, Vol. 14, eds. D.J. Ernst, I.E. Perakis, and A.S. Umar, (Nova Science Publishers, Huntington NY, 2000) p. 85.

[33] L.S. Brown, *Am. J. Phys.* **41**, 525, (1973).

[34] M. Nauenberg, *Phys. Rev.* **A 40**, 1133, (1989).

[35] A. Buchleitner, D. Delande, and J. Zakrzewski, *Phys. Rep.* **368**, 409, (2002).

[36] D. Vrinceanu, and M.R. Flannery, *Phys. Rev.* **A 60**, 1053, (1999).

[37] M.R. Flannery, and D. Vrinceanu, *Phys. Rev.* **A 65**, 022703, (2002).

[38] L.I. Schiff, *Quantum Mechanics*, 3rd ed. (McGraw-Hill, New York, 1968).

[39] D. Comparet, T. Vogt, N. Zahzam, M. Mudrich, and P. Pillet, *Mon. Not. R. Astron. Soc.* **361**, 1227, (2004).

[40] P. Mansbach and J. Keck, *Phys. Rev.* **181**, 275, (1969).

[41] H. Reinholz, R. Redmer, G. Röpke, and A. Wierling, *Phys. Rev.* **E 62**, 5648, (2000).

[42] E.W. Schmid, H. Ziegelmann, *The Quantum Mechanical Three-Body Problem* (Vieweg,

Braunschweig, 1974).

[43] J.P. Coleman, in *Case studies in Atomic Collision Physics I*, eds. E.W. McDaniel and M.R.C. McDowel (North-Holland, Amsterdam, 1969) p. 99.

[44] A. Könies and S. Günter, *Phys. Rev.* **E 52**, 6658, (1995).

In: Advances in Plasma Physics Research, Volume 7
Editor: Francois Gerard
ISBN: 978-1-61122-983-7

Chapter 3

PLASMA TURBULENT TRANSPORT MODELLING BY MEANS OF LÉVY DISTRIBUTIONS

R. Sánchez[1],[*] B.Ph. van Milligen[2] and B.A. Carreras[3]
[1]Departamento de Física, Universidad Carlos III de Madrid,
28911 Leganés, Madrid, SPAIN
[2]Laboratorio Nacional de Fusión, Asociación EURATOM-CIEMAT,
28040 Madrid, SPAIN
[3]Fusion Energy Division, Oak Ridge National Laboratory,
Oak Ridge, TN 37831, U.S.A.

Abstract

The experimental study of transport in tokamaks has revealed a number of *strange phenomena*, apparently related to the fact that the plasma profiles are close to some critical threshold: Bohm scaling of the energy confinement time, stiff profiles, on-axis/off-axis heating paradoxes, superdiffusive pulse propagation, and others. These observations have cast some doubt on the appropriateness of the current description of transport, based on diffusivities and conductivities. In particular, the extrapolation of analysis results to larger devices may depend critically on this issue, with possibly important consequences for the design of next-generation tokamaks such as ITER.

In this paper, we will argue that the modelling of these strange phenomena requires a generalization of the usual diffusive transport paradigm on a fundamental level. In order to do so, recall that typical particle diffusion is the macroscopic consequence of the microscopic random motion of individual particles, described by a probability distribution function (pdf) with finite moments. This fact, combined with the central limit theorem that asserts that any sum of random variables obeying pdfs with finite moment converges to a Gaussian, yields the standard diffusive picture. However, Gaussians are not the only limit distributions produced by the central limit theorem: the larger family of Lévy distributions is obtained when pdfs with divergent moments are considered instead.

To illustrate these ideas, we will discuss the construction of several simple models in which the random particle motion is described by Lévy distributions, governed by

---

[*]E-mail address: sanchezferlr@ornl.gov. Corresponding author. Present address: Fusion Energy Division, Oak Ridge National Laboratory, P.O. Box 2008, Oak Ridge, TN 37831-6961, USA.

a critical gradient. The evolution equation for the particle density can then obtained in the form of a Master Equation or, by taking the fluid limit, of a Fractional Differential Equation. In spite of their simplicity, these models naturally produce the rapid propagation of pulses, Bohm-like scaling, and even on-axis peaking of the density in the presence of off-axis sources, similar to what is observed in actual tokamaks and stellarators. More importantly, this new approach shows that diffusivities and conductivities measured in relatively small devices may lead to error when extrapolated directly to predict transport in larger devices, and teaches us which physical quantities should be used instead to ensure that transport is properly scaled.

1 Introduction

One of the most promising methods to produce net energy from thermonuclear fusion in an economically viable way is the magnetic confinement of plasmas in toroidal configurations, such as the tokamak [1] or the stellarator [2]. However, a plasma confined in any of these devices is forced to stay very far from thermodynamical equilibrium, as is evident from the enormous density and temperature gradients that need to be sustained to produce fusion conditions. Spontaneously, the plasma will attempt to reduce these gradients by inducing particle and energy fluxes directed radially out of the machine [1]. These radial fluxes limit the maximum time that the plasma can be confined. Their understanding and control is, therefore, essential to the success of this approach [5].

The usual theoretical approach to understanding these radial fluxes is well illustrated by neoclassical plasma transport theory [6, 7]. In this approach, all radial fluxes are obtained in terms of a "surface-averaged" transport matrix M, that relates the radial fluxes (i.e., the *particle flux*, Γ, and the *electron and ion energy* fluxes, $q_{e,i}$) *locally* to the thermodynamic forces that induce them (*temperature* and *pressure gradients*, *electric fields*), under the assumption that dissipation is mainly due to collisional Coulomb scattering, while proximity to equilibrium is assumed, thus enabling the application of linear techniques, such as the Chapman-Enskog scheme [8].

The same formal structure of transport equations has been routinely adopted (in simplified form) to try to model radial transport when **turbulence** is also considered. Turbulence is needed to account for the strong **anomalous** component (i.e., beyond the neoclassical prediction) that dominates in almost all experiments [9]. Using different strategies to perform turbulent renormalization for particular kinds of unstable modes, most theoretical models of turbulence aim at estimating turbulence-enhanced values for the relevant diffusivities and conductivities, without modifying the formal modelling framework. A typical example is provided by the Weiland model for drift-wave induced radial transport (r is the radial coordinate) [10]:

$$\frac{\partial}{\partial t}\begin{pmatrix}T_i\\T_e\\n\end{pmatrix}=\frac{1}{r}\frac{\partial}{\partial r}\left[r\begin{pmatrix}\chi_{ii}&\chi_{ie}&\chi_{in}\\0&\chi_{ee}&\chi_{en}\\0&D_{ne}&D_{nn}\end{pmatrix}\frac{\partial}{\partial r}\begin{pmatrix}T_i\\T_e\\n\end{pmatrix}\right]+\begin{pmatrix}S_i\\S_e\\S_n\end{pmatrix}+R\begin{pmatrix}T_e-T_i\\T_i-T_e\\0\end{pmatrix} \quad (1)$$

[1]Of course, these are not the only fluxes induced. Fluxes in other directions, including momentum fluxes, will also appear.

that describes the radial transport of the ion and electron temperature and the plasma density in the presence of sources (the S_k vector) and friction between electrons and ions (the R term). The local relation between fluxes and forces is reflected in the transport matrix that contains conductivities, diffusivities and cross-coupling terms derived from a drift-wave turbulent model [10]. Other, similar approaches use more complex equations that combine neoclassical theory results and/or quasilinear results from gyrokinetic, gyrofluid and other approaches to turbulent transport [11, 12, 13, 14, 15, 16, 17, 18].

However, the assumption that fluxes and forces are related *locally*, in the sense that fluxes depend only on the value of the forces at the same point, may not be fully justified. Many experimental results obtained in the last ten to twenty years have shown that our understanding of these fluxes is poor at best. For instance, the scaling of the **energy confinement time** defined by:

$$\tau_E \equiv W_E \cdot \left(P_{ext} - \frac{dW_E}{dt} \right)^{-1}, \tag{2}$$

where W_E is the stored plasma energy and P_{ext} is the external power supplied to the confined plasma, has remained a mistery. It has systematically been observed that:

$$\tau_E \propto P_{ext}^{-(0.7\pm0.1)}, \tag{3}$$

a phenomenon known as **power degradation** [3, 4, 5]. This result contradicts the expectation from the local approach, since the latter would predict that the confined energy be linearly proportional to the supplied external power.

Other relevant experimental findings in this regard are provided by the observation of **canonical profiles** [19]. In many experiments, it is observed that the steady state density and temperature profiles are rather insensitive to the strength and spatial distribution of the external fueling and power input. Moreover, if the source is located at an off-center radial location, sometimes the same profile is observed that would be obtained with a centered source, i.e. peaked at the center. One deduces that **uphill transport** seems to take place in the tokamak (at least, in L-mode), which is again apparently in contradiction with the standard paradigm [20]. To account for this phenomenon, fast radial pinches must be invoked, directed up the gradient, for which usually no clear physical origin can be found, and that seem to be different in each experiment [21].

Probably the most convincing evidence of the failure of the so-called standard transport paradigm is obtained in *perturbative experiments* [22, 23, 24, 25, 26, 27, 28]. Here, the temporal evolution and spatial propagation of a small perturbation of the system steady state is observed. Due to the small size of the perturbation, a local (and even linear) approach would seem to be well justified, in spite of the fact that the steady state is very far from equilibrium. However, in many perturbative experiments, it has been observed that the perturbation propagates **superdiffusively** (and sometimes even ballistically), at speeds that exceed the values expected from the appropriate characteristic diffusive velocity considerably. In some cases, the plasma core seems to react right after the excitation of the perturbation has been produced at another location, such that the latter apparently has not had enough time to propagate to the core.

In the last few years, an important theoretical effort has been devoted to the explanation of the possible origin of these "non-local" transport effects. On the one hand, it was soon pointed out that the observation of canonical profiles and power degradation might be related to the existence of some (local) **critical gradient** above which a fast transport channel might become active [29, 30]. On the other, several ideas have also been proposed regarding the nature of this fast channel. From standard turbulence theory, concepts such as **"streamers"** (vortex-like, radially elongated structures developed in temperature gradient driven turbulence that can enhance radial transport) have been proposed [31, 32]. From a more basic perspective, ideas based on the concept of **self-organized criticality** (SOC) [33] have been advanced, in which the non-local transport events would correspond to so-called **"avalanches"**. These avalanches would correspond to the successive destabilization of radially localized instabilities, interacting nonlinearly to yield a transport channel with no characteristic radial scale lengths [34, 35, 36, 37]. And indeed, a certain amount of experimental evidence has been found that might support these ideas to some extent [38, 39, 40, 41, 42].

It is not clear, however, whether any local formal framework such as that provided by Eq. 1 would be adequate to include such "non-local" elements. The meaningfulness of diffusivities and conductivities relies on the existence of well-defined characteristic lengths and times at a microscopic scale. Consider, e.g., the Larmor radius and the inverse collision frequency in the case of neoclassical transport, and the radial coherence length and the decorrelation time for turbulence. In contrast, the maximum radial lengths of avalanches or streamers would be limited only by the system size (set by the tokamak minor radius), and would thus lack any characteristic length, since their average size would diverge (grow) with the system size.

In this article, we will try to explore whether the basic ideas underlying Eq. 1 can be extended so as to overcome some of the aforementioned limitations. For simplicity, we will consider an imaginary system in which only one field can be transported: the particle density. In that case, and ignoring geometric considerations in what follows, Eq. 1 would reduce to the classical diffusive equation:

$$\frac{\partial n}{\partial x} = \frac{\partial}{\partial x}\left(D(x)\frac{\partial n}{\partial x}\right) + S_n(x). \tag{4}$$

As we will show, at least one such extension is possible for this one-field model that is based on the so-called **Lévy distributions** [43, 44]. Many of the previously discussed phenomena might be well captured by this formalism, with the added beauty of containing Eq. 4 as a particular case: namely, when characteristic scales are indeed present in the problem.

2   Continuous Time Random Walks: Microscopic Origin of the Diffusion Equation

To understand how the transport equation Eq. 4 can be extended to accommodate non-local transport mechanisms, we must first review a rather old concept: the **continuous time random walk** (CTRW). CTRWs have found a wide range of applications in physics since their introduction almost forty years ago [45]. These generalizations of the standard (discrete) random walk are defined in terms of two probability density functions (pdfs): a **step size pdf** $p(x - x')$, giving the probability of a walker moving from x' to x at time t, and a **waiting time pdf** $\psi(t - t')$, giving the probability of having waited an amount of time $t - t'$ at x' before moving to x.

Such a CTRW is readily *"integrated"*. That is, it is possible to derive a formal expression for the probability of the walker being at x at time t, $n(x, t)$. This quantity is also referred to as the *walker density*[2]. The derivation exploits the spatial invariance of $p(x - x')$ and the temporal invariance of $\psi(t - t')$, to solve for the Laplace-Fourier transform of the walker density formally[3], which happens to be [45, 46, 47]:

$$n(k, s) = n_0(k)[1 - \psi(s)] \left[s\left(1 - \psi(s)p(k)\right) \right]^{-1}, \tag{5}$$

where $n_0(k)$ is the Fourier transform of the initial walker density. Eq. 5, also known as the **Montroll-Weiss equation**, can then be Laplace-Fourier inverted to complete the integration and provide the walker density for all t and x.

It is also straightforward to prove that this CTRW can be mapped to the following **Generalized Master Equation** (GME) [48, 49, 50]:

$$\frac{\partial n(x, t)}{\partial t} = \int_0^t dt' \left[\int_{-\infty}^{+\infty} dx' K(x - x', t - t')n(x', t') - n(x, t') \int_{-\infty}^{+\infty} dx' K(x - x', t - t') \right], \tag{6}$$

provided the GME transition kernel K is chosen as:

$$K(k, s) = s\psi(s)p(k)\left[1 - \psi(s)\right]^{-1}. \tag{7}$$

Again, the spatial and temporal invariance of the problem has been exploited.

This CTRW/GME formalism can be used to model the transport properties of many systems, once appropriate forms for p and ψ are chosen that capture the relevant physics of the mechanism that governs transport. Thanks to the central limit theorem [51], in many cases the pdfs will converge to an **exponential** for ψ, and a **Gaussian** for p. The "**fluid limit**" of the CTRW, in which only those details pertinent to the long-time, long-distance system dynamics are kept (in Eq. 5 or 7), then yields the classical diffusive equation [52, 53]:

$$\frac{\partial n}{\partial t} = D\frac{\partial^2 n}{\partial x^2}, \tag{8}$$

where the diffusion coefficient D is determined uniquely by the characteristic parameters that define the chosen distributions[4].

---

[2]The reason for the name is that, if a population of walkers is started simultaneously, while their motion is governed by the same pdf, $n(x, t)$ would give their distribution at any later time.

[3]In what follows, any quantity and its Fourier and/or Laplace transforms will be represented by the same symbol, since they can still be distinguished by their arguments. s and k are used respectively as Laplace and Fourier variables.

[4]This equation is the infinite-domain analog to Eq. 4. When constrained to a finite domain, a source must be also added so that steady state solutions may exist.

To see how this comes about, let's consider the following step size and waiting time pdfs:

$$\psi(t - t') = \tau_1^{-1} \cdot \exp\left(-\frac{t - t'}{\tau_1}\right), \quad p(x - x') = \frac{1}{2\sigma_2\sqrt{\pi}} \exp\left(-\frac{(x - x')^2}{4\sigma_2^2}\right). \quad (9)$$

Note that, in this case, the walker motion has *well defined characteristic length* (given by σ_2) and *time (τ_1) scales*[5]. To take the fluid limit, we take the $s \to 0$ limit of the waiting time pdf and the $k \to 0$ limit of the step size pdf,

$$\psi(s) \simeq 1 - \tau_1 s, \quad p(k) \simeq 1 - \sigma_2^2 k^2, \quad (10)$$

and insert these into Eq. 7 for the GME transition kernel:

$$K(s, k) \simeq \tau_1^{-1}(1 - \tau_1 s)(1 - \sigma_2^2 k^2). \quad (11)$$

Inserting this expression into the Fourier-Laplace transform of Eq. 6,

$$sn(s, k) - n_0(x) = [K(s, k) - K(s, 0)] \, n(s, k), \quad (12)$$

and keeping only the lowest-order terms in s and k, we obtain:

$$sn(s, k) - n_0(x) \simeq -\left(\frac{\sigma_2^2}{\tau_1}\right) k^2 n(s, k), \quad (13)$$

which can easily be Laplace-Fourier inverted to yield Eq. 8. It also gives the value of the diffusive coefficient:

$$D \equiv \frac{\sigma_2^2}{\tau_1}, \quad (14)$$

which, as stated previously, is *fully determined* by the characteristic length and time scales of the walker motion, prescribed via the choice for the step size and waiting time pdfs [6].

3 Continuous Time Random Walks: beyond the Diffusion Equation

But the applicability of integrable CTRWs surpasses standard diffusive systems. Experiments have shown that, in many systems of physical, chemical and biological interest, the variance of the walker displacement from some initial point increases with time as [47, 52, 53, 54, 55, 56, 57, 59, 60]:

$$\left\langle |x - x_0|^2 \right\rangle \propto t^\nu, \quad \nu \neq 1, \quad (16)$$

---

[5]Characteristic in the sense that they correspond to the first nonvanishing moments of the distributions: the second, for the Gaussian and the first, for the exponential. Any higher moment is a function of these.

[6]Note that the usual formula,

$$D = \frac{(\Delta x)^2}{2\Delta t} \quad (15)$$

comes about because the usual definition of the Gaussian width Δx relates to σ through $\Delta x = \sqrt{2}\sigma$.

in contrast to what Eq. 8 predicts ($\nu = 1$) [53]. In these cases, transport is termed either "superdiffusive" ($\nu > 1$) or "subdiffusive" ($\nu < 1$)[7]. Integrable CTRWs can still be used to describe transport in many of these cases, but ψ and p should now be chosen (with certain restrictions that will be made precise later) from the family of **stable Lévy distributions** (see Appendix A). This family of pdfs, usually denoted by $P_{[\alpha,\beta,\sigma]}(y)$, satisfies a generalized version of the central limit theorem that *does not require that the pdfs decay exponentially at large values of the argument* [43, 44]. It contains, as special cases, the Gaussian distribution for $\alpha = 2, \beta = 0$, and the exponential distribution for $\alpha = 1, \beta = 1$ (the latter is not strictly contained, but exists as a weak limit when $\alpha \to 1$ for $\beta = 1$ [61]). But the more interesting property of these distributions is that they *lack characteristic scales*, since *any moment of the distribution of order higher than α diverges for alpha < 2*. For this reason, they are extremely useful to model transport in those cases where transport characteristics diverge with system size, thus providing the natural formalism to model systems where non-local transport effects are present.

A "fluid limit" also exists for these CTRWs, expressed in terms of **fractional differential operators** [52, 53]. In spite of the somewhat esoteric nature of these operators (see Appendix B), efficient algorithms exist that can be used to integrate them numerically for most applications [62, 63, 64, 65, 66]. At this stage, the only thing we need to know about Lévy distributions is that they can be defined in terms of their Fourier transform [44]:

$$P_{\alpha,\beta,\sigma}(k) = \exp\left[-\sigma^\alpha |k|^\alpha \left(1 - i\beta \mathrm{sgn}(k)\tan\left(\frac{\pi\alpha}{2}\right)\right)\right], \qquad (17)$$

with $\alpha \in 0,2$, $|\beta| \leq 1$ and $0 < \sigma < \infty$ (the meaning of each label being discussed in Appendix A). Usually, one can choose any stable Lévy pdf as step size pdf (note that the choice $\alpha = 2, \beta = 0$ is the Gaussian pdf), but waiting time pdfs can only be defined for positive lapses of time (i.e., for $t - t' \geq 0$). For this reason, they must be chosen within the subfamily of Lévy pdfs known as *positive extremal distributions* ($\alpha < 1, \beta = 1$) [44], that are only defined for positive values of y (see Appendix A). Also, the exponential pdf can be used, since it can be shown that it is the limiting pdf when the limit $\alpha \to 1$ for $\beta = 1$ is taken [61].

Next, it is useful to introduce some notation. In what follows, the labels α, β, σ will always refer to step size pdfs. For the waiting time pdfs, only α and σ are free parameters, since $\beta = 1$. To avoid confusion with the step size labels, we will instead use γ (for α) and τ (for σ) when referring to waiting time distributions. Therefore, we will assume that the integrable CTRW is defined by a waiting time step size:

$$\psi(t - t') = P_{[\gamma,1,\tau]}(t - t'), \quad \gamma \leq 1, \ 0 < \tau < \infty, \qquad (18)$$

and step size pdf:

$$p(x - x') = P_{[\alpha,\beta,\sigma]}(x - x'), \quad \alpha \leq 2, \ |\beta| \leq 1, \ 0 < \sigma < \infty. \qquad (19)$$

---

[7]Note that the fast propagation of pulses observed in tokamaks and stellarators would correspond to some kind of superdiffusive propagation.

The fluid limit can now be taken, for instance, of the Montroll-Weiss equation (Eq. 5). We only need to take the limit of long distances (in Fourier space, $k \to 0$) in $p(k)$ and of long times (in Laplace space, $s \to 0$) in $\psi(s)$. This reduces to approximating Eq. 17 as:

$$p(k) \simeq 1 - \sigma^\alpha |k|^\alpha \left(1 - i\beta \mathrm{sgn}(k) \tan\left(\frac{\pi\alpha}{2}\right)\right), \tag{20}$$

and approximating the Laplace transform of positive extremal Lévy pdfs, given by Eq. 73, by:

$$\psi(s) \simeq 1 - A_\gamma^{-1} \tau^\gamma s^\gamma. \tag{21}$$

where we have also included the exponential law when $\gamma = 1$ and defined the constant:

$$A_\gamma = \begin{cases} \cos\left(\frac{\pi\gamma}{2}\right), & \gamma < 1 \\ \\ 1, & \gamma = 1 \end{cases} \tag{22}$$

After inserting Eqs. 21 and 20 in Eq. 5, the fluid limit of the Montroll-Weiss equation becomes:

$$n(s,k) \simeq n_0(k) \left[s + C(\alpha,\gamma)s^{1-\gamma}|k|^\alpha \left(1 - i\beta\, \mathrm{sgn}(k) \tan\left(\frac{\pi\alpha}{2}\right)\right)\right]^{-1}. \tag{23}$$

where the coefficient $C(\alpha,\gamma) = A_\gamma \sigma^\alpha / \tau^\gamma$ has been defined. Eq. 23 can be rewritten as:

$$sn(s,k) - n_0(k) = -C(\alpha,\gamma)s^{1-\gamma}|k|^\alpha \left(1 - i\beta\, \mathrm{sgn}(k) \tan\left(\frac{\pi\alpha}{2}\right)\right) n(s,k). \tag{24}$$

After using the identity Eq. 84, Eq. 24 can be Fourier inverted by introducing the two **Riemann-Liouville fractional differential operators** defined by Eq. 79, which satisfy [62, 64]:

$$\mathsf{F}\left[\frac{\partial^\alpha n}{\partial(\pm x)^\alpha}\right] \equiv (\mp ik)^\alpha n(k). \tag{25}$$

$\mathsf{F}[\cdot]$ represents the Fourier transform. The resulting equation thus becomes a fractional differential equation (FDE) in space:

$$sn(s,x) - n_0(x) = -\frac{C(\alpha,\gamma)s^{1-\gamma}}{2\cos\left(\frac{\pi\alpha}{2}\right)}\left[(1+\beta)\frac{\partial^\alpha n}{\partial x^\alpha} + (1-\beta)\frac{\partial^\alpha n}{\partial(-x)^\alpha}\right]. \tag{26}$$

In order to carry out the Laplace inversion of Eq. 26, two choices are possible. The first one is to multiply both sides by $s^{\gamma-1}$ and introduce the **Caputo fractional differential operator** [67] (Eq. 85), which verifies [62, 64] (for $\gamma < 1$):

$$\mathsf{L}\left[\frac{\partial_c^\gamma n}{\partial_c t^\gamma}\right] \equiv s^\gamma n(s,x) - s^{\gamma-1}n_0(x), \tag{27}$$

where $\mathsf{L}[\cdot]$ represents the Laplace transform. The result is the FDE in space and time:

$$\frac{\partial_c^\gamma n}{\partial t_c^\gamma} = -\frac{C(\alpha,\gamma)}{2\cos\left(\frac{\pi\alpha}{2}\right)}\left[(1+\beta)\frac{\partial^\alpha n}{\partial x^\alpha} + (1-\beta)\frac{\partial^\alpha n}{\partial(-x)^\alpha}\right]. \tag{28}$$

A second possibility is to Laplace invert Eq. 26 directly. This can be done by introducing the **Riemann-Liouville differential operator with start point at** $t = 0$ (Eq. 78) [68]:

$$\frac{\partial n}{\partial t} = - {}_0D_t^{1-\gamma} \left\{ \frac{C(\alpha,\gamma)}{2\cos\left(\frac{\pi\alpha}{2}\right)} \left[(1+\beta)\frac{\partial^\alpha n}{\partial x^\alpha} + (1-\beta)\frac{\partial^\alpha n}{\partial(-x)^\alpha}\right] \right\}. \tag{29}$$

3.1 Interpretation of FDEs Eqs. 28 and 29.

The interpretation and applications of Eqs. 28 and 29 have been discussed in detail in the literature for different choices of α and γ [50, 52, 53, 68, 69, 70]. One interesting remark is that the exponent that determines the superdiffusive or subdiffusive character of transport is equal to $\nu = 2\gamma/\alpha$ (see Eq. 16). Thus, superdiffusive behaviour is observed when $2\gamma > \alpha$, diffusive behaviour when $2\gamma = \alpha$ and subdiffusive behaviour when $2\gamma < \alpha$ [52, 53]. For the choices $\alpha = 2, \gamma = 1$, Eqs. 28 and 29 reduce to the standard diffusive equation (see Eq. 8) with diffusive coefficient $D = C(2,1) = \sigma^2/\tau$.

We proceed now to interpret each term in the fluid limit given by Eq. 29. To do so, it is convenient to set $\gamma = 1$ for the moment, and consider the Markovian version of Eq. 29:

$$\frac{\partial n(x,t)}{\partial t} = - \left\{ \frac{C(\alpha,\gamma)}{2\cos\left(\frac{\pi\alpha}{2}\right)} \left[(1+\beta)\frac{\partial^\alpha n}{\partial x^\alpha} + (1-\beta)\frac{\partial^\alpha n}{\partial(-x)^\alpha}\right] \right\}. \tag{30}$$

The r.h.s. of Eq. 30 contains two terms. The first one, proportional to $1 + \beta$, is the only one that survives if $\beta = 1$ [8]. However, the α-fractional derivative is nothing but an integral over $(-\infty, x]$ (see Eq. 78):

$$\frac{\partial^\alpha \Lambda_1^{(1)}}{\partial x^\alpha} \equiv \frac{1}{\Gamma(p-\alpha)} \frac{d^p}{dx^p} \int_{-\infty}^{x} \frac{n(x',t)dx'}{(x-x')^{\alpha-p+1}}, \tag{31}$$

where p is the integer part of α. Therefore, this fractional derivative collects the contributions from all walkers that end up at x at time t *from* $x' \le x$. The term is thus intrinsically **non-local**!

Analogously, the second term, proportional to $(1 - \beta)$, gives the contribution to the change in $n(x,t)$ from points with $x' \ge x$. In the general case, a combination of both terms applies [52, 53]. A particular case is that in which the combination of the two contributions yields a symmetric Lévy pdf: each walker has equal probability of moving to larger or smaller x's from any given location. The resulting equation can be written in a familiar compact form using the Riesz operator (Eq. 82, Appendix B):

$$\frac{\partial n(k,t)}{\partial t} = C(\alpha,\gamma)\frac{\partial^\alpha n}{\partial|x|^\alpha}, \tag{32}$$

---

[8]Note that, at the "microscopic" level, $\beta = 1$ corresponds to having the walker moving according to a step size Lévy pdf in which the only steps allowed are those that take the walker to larger x's (except for an exponentially vanishing contribution to lower x's).

that clearly reduces to the classical diffusive equation for $\alpha = 2$.

Now, let us consider the non-Markovian case with $\gamma < 1$, that allows modelling memory effects in a probabilistic manner, associated with the "microscopic" waiting time pdf ψ. Writing the fractional time derivative explicitly, it happens that (recall Eq. 78, Appendix B):

$$ {_0}D_t^{1-\gamma} \, n(x', t) = \frac{1}{\Gamma(m - 1 + \gamma)} \frac{d^m}{dt^m} \left[\int_0^t \frac{n(x', t')}{(t - t')^{2-\gamma-m}} dt' \right], $$

where m is the integer part of $1 - \gamma$. This operation thus collects contributions from all past times $t' < t$, providing a "memory" of the *previous history of the system*.

4 Nonintegrable Continuous Time Random Walks: a Formalism for Plasma Turbulent Transport?

Above, we have shown that the GME/CTRW provides a very powerful framework to model both non-local and non-Markovian transport processes, when either superdiffusive or subdiffusive propagation is observed. This suggests that such a framework might be extremely useful to model radial transport in magnetically confined plasmas [57, 58]. But a central assumption of the integrable GME/CTRW theory is that the system must be both spatially and temporally invariant: *every location is probabilistically identical at any time*. Regretfully, the experimental picture gathered in magnetically confined plasmas indicates that spatial invariance may be absent in these systems. Indeed, power degradation and the existence of canonical profiles suggest that a fast transport mechanism exists that becomes active *only at those points* where some (usually nonlinear) threshold condition is overcome [21, 71]. Linear stability theory also supports this conclusion, since many instabilities are known that become unstable when a critical pressure or temperature gradient is exceeded [72]. Turbulent simulations of these modes have shown that the corresponding transport takes place through the excitation of "avalanche"-like radial events [36, 73, 74]. Furthermore, if statistics of the advection of tracer particles following the $\vec{E} \times \vec{B}$ (i.e. electrostatic) flow excited by these modes are collected, it is found that their transport is strongly non-diffusive and non-local, apparently well described in terms of Lévy distributions [75, 76, 77].

The inadequacy of integrable CTRWs in a situation like this can also made apparent in the context of the simple one-field model. For instance, let's assume that a fast, non-local transport channel becomes active when a critical density gradient is locally overcome, while transport is dominated by a slower, local channel otherwise. This situation can be easily modeled by choosing the following step size pdf:

$$ p(x - x', x', t) = \lambda_1(x', t) P_{[\alpha, \beta, \sigma]}(x - x') + \lambda_2(x', t) P_{2, 0, \sigma'}(x - x'), \tag{33} $$

where we have characterized the fast process by a Lévy pdf with $\alpha < 2$, while the slow process is diffusive and characterized by a Gaussian. The two "projectors" λ_1, λ_2 may then be defined as:

$$ \lambda_1(x, t) = H\left(\left| \frac{dn}{dx}(x, t) \right| - Z_c(x) \right), \quad \lambda_2 = 1 - \lambda_1 \tag{34} $$

with $H(x)$ the usual Heaviside step function and $Z_c(x)$ the local critical gradient. Clearly, the step size pdf describing this situation has lost its spatial invariance (due to its dependence on x'), preventing any Fourier transform technique to be applicable in this case, so that no equation analogous to the Montroll-Weiss equation (Eq. 5) can be derived.

Therefore, if the GME/CTRW framework is to be useful to describe transport in these cases, it must be extended to include at least spatially non-invariant cases similar to that given by Eq. 33. This can be done by observing that *the class of CTRWs that can be mapped to a GME is larger than that of the integrable CTRWs.* It also includes all nonintegrable CTRWs with a step size pdf of the form $p(x - x', h(x', t))$, where h is any arbitrary (nonlinear) function of the form [78, 79]:

$$h(x', t) = f\left(x', t; n(x', t), \frac{dn}{dx}(x', t), \frac{d^2 n}{dx^2}(x', t) \cdots\right). \tag{35}$$

Therefore, the step size pdf may contain arbitrary non-linearities and/or inhomogeneities via the function h, thus permitting the next step that a walker makes at time t to depend non-linearly, but in a Markovian way, on any local quantity (i.e., defined at x' and t).

4.1   Derivation of a GME for Nonintegrable CTRWs

The difficulties of proving that an GME can be associated to this CTRW become clear when trying to "integrate" it along the lines outlined in Secs. 2 and 3. First, we express the probability of finding the walker as [46]:

$$n(x, t) = \int_0^t \eta(x; t - t') Q(x; t') dt', \tag{36}$$

where $\eta(x; t - t')$ represents the probability that the walker, located at x' at time t', still remains in the same position at time t:

$$\eta(x, \tau) = \int_0^\tau d\tau' \, \psi(\tau', x). \tag{37}$$

$Q(x; t)$ represents the total probability of the walker arriving at position x at time t by any possible route. Next, we Laplace transform Eq. 36 to get,

$$n(x, s) = \eta(x; s) Q(x; s). \tag{38}$$

The Laplace transform of $\eta(x, t - t')$ is trivially obtained in terms of $\psi(s, x)$ by Laplace transforming Eq. 37:

$$s\eta(x, s) = 1 - \psi(s, x). \tag{39}$$

Regarding $Q(x; t)$, it satisfies the recursive equation as shown in Ref. [46],

$$Q(x; t) - \delta(x)\delta(t) = \int_{-\infty}^{\infty} dx' p[x - x', h(x'; t)] \int_0^t dt' \psi(x'; t - t') Q(x'; t'), \tag{40}$$

that only assumes that the walker is initially located at x. From Eq. 40 it is clear that, if p and ψ would not depend explicitly on h, $Q(q,s)$ would be readily available via a Fourier-Laplace transformation. However, in the case of our CTRW, $Q(x,t)$ may depend on $n(x',t)$ (through the function h), and the standard approach from Ref. [46] is no longer applicable. The CTRW under consideration is therefore *"nonintegrable"* due to the presence of the nonlinearity.

To derive the GME associated with this CTRW we must establish the link to the CTRW through Eq. 40 instead [78, 79]. We start by introducing an auxiliary function in Laplace space,

$$\phi(x;s) = \psi(x;s)/\eta(x;s), \tag{41}$$

that allows us to rewrite Eq. 40 as:

$$Q(x;t) - \delta(x)\delta(t) = \int_{-\infty}^{\infty} dx' p\left[x - x', h(x';t)\right] \int_0^t dt' \phi(x';t-t')n(x',t'), \tag{42}$$

after transforming the temporal convolution in the r.h.s. of Eq. 40 with the help of the Laplace transform $\mathsf{L}[\cdot]$:

$$\mathsf{L}\left[\int_0^t dt' \psi(x';t-t')Q(x';t')\right] = \psi(x',s)Q(x',s) =$$

$$= \phi(x',s)n(x',s) = \mathsf{L}\left[\int_0^t dt' \phi(x';t-t')n(x';t')\right], \tag{43}$$

where we have also used Eq. 38. Next, we Laplace-transform Eq. 42, multiply the result by $s\eta(x;s)$ and use Eq. 38 (after adding and subtracting $\delta(x)$) to obtain:

$$[sn(x,s) - \delta(x)] - \delta(x)[s\eta(x;s) - 1] = s\eta(x;s)g(x;s), \tag{44}$$

where $g(x,s)$ stands for the Laplace transform of the r.h.s. of Eq. 42. $g(x,s)$ is eliminated by combining the Laplace-transform of Eq. 42 with Eqs. 39 and 41 to get:

$$g(x;s) = \frac{\delta(x)\psi(x;s) - \phi(x;s)n(x,s)}{(s\eta(x;s) - 1)}. \tag{45}$$

After inserting this expression for $g(x,s)$, Eq. 44 is Laplace-inverted to yield the final GME we sought:

$$\frac{\partial n(x,t)}{\partial t} = -\int_0^t dt' \phi(x;t-t')n(x,t') +$$

$$+ \int_0^t dt' \int_{-\infty}^{\infty} dx' \phi(x';t-t')p\left[x - x', h(x';t)\right]n(x',t'). \tag{46}$$

The resulting GME transition kernel in Eq. 46 is thus,

$$K(x,x',t,t') = \phi(x';t-t')p\left[x - x', h(x';t)\right], \tag{47}$$

that reduces to the usual transition kernel given by Eq. 7 if spatial invariance is again assumed by disregarding any possible dependence on $h(x',t)$. The function $\phi(t-t')$ is usually known as the *memory function* since it becomes a delta function only when the CTRW is Markovian [80]. As we will see, it is closely related to the fractional time derivatives that we already encountered in Sec. 3.

4.2   Fluid limit of Nonintegrable CTRWs

Analogously to what we did in the case of integrable CTRWs, it is also possible to derive the fluid limit of the GME Eq. 46 if the choices of waiting time and step size pdfs are those suggested by the generalized central limit theorem: Lèvy distributions [81, 82]. Regarding the waiting time pdf, the same restrictions apply as with the integrable case (recall Eq. 18). However, for the step size pdf we will consider the combination of two arbitrary stable Lévy pdfs[9], and inspired by our previous discussion we write:

$$p(x - x', x', t) = \lambda_1(x', t)P_{[\alpha_1,\beta_1,\sigma_1]}(x - x') + \lambda_2(x', t)P_{[\alpha_2,\beta_2,\sigma_2]}(x - x'). \tag{48}$$

The choice for the projectors λ_1, λ_2 is kept completely arbitrary, as long as they satisfy:

$$\lambda_1(x, t) + \lambda_2(x, t) = 1, \quad \forall x, t, \tag{49}$$

and are positive everywhere.

We proceed by first taking the fluid limit ($s \to 0$) of the memory function:

$$\phi(s) = \frac{s\psi(s)}{1 - \psi(s)} \sim A_\gamma \tau^{-\gamma} s^{1-\gamma} \tag{50}$$

for which only Eq. 21 is required (the coefficient A_γ was introduced in Eq. 22 in Sec. 2). We will use this result to take the fluid limit of the first term on the r.h.s. of GME Eq. 46, by introducing again the Caputo derivative [67] and taking advantage of Eq. 27, so that we obtain:

$$\int_0^t dt' \phi(x; t - t')n(x, t') = \mathsf{L}^{-1}\left[\phi(s)n(x, s)\right] \simeq \mathsf{L}^{-1}\left[A_\gamma\tau^{-\gamma}s^{1-\gamma}n(x, s)\right] =$$

$$= A_\gamma\tau^{-\gamma}\left[\frac{\partial_c^{1-\gamma}n}{\partial_c t^{1-\gamma}} + \frac{t^{\gamma-1}n_0(x)}{\Gamma(\gamma)}\right] = A_\gamma\tau^{-\gamma}\left[{}_0D_t^{1-\gamma}n\right]. \tag{51}$$

To derive this expression, use has also been made of [64]:

$$L\left[t^\gamma\right] = \Gamma(\gamma + 1)s^{-(\gamma+1)}, \tag{52}$$

and of the relation between the Caputo derivative and the Riemann-Liouville derivative with start point at $t = 0$ (Eq. 87, in Appendix B). Doing the same with the time convolution appearing inside the second term of the r.h.s. of Eq. 46, we can rewrite the GME as:

$$\frac{\partial n(x, t)}{\partial t} = -A_\gamma\tau^{-\gamma}\left\{\left[{}_0D_t^{1-\gamma}n\right](x, t) + \right.$$

$$\left. + \int_{-\infty}^\infty dx' p\left[x - x', h(x'; t)\right]\left[{}_0D_t^{1-\gamma}n\right](x, t)\right\}. \tag{53}$$

Next, we take the spatial part of the fluid limit by computing the Fourier transform of Eq. 53 and taking its limit when $k \to 0$:

$$\frac{\partial n(k, t)}{\partial t} = -\sum_{j=1}^2 C(\alpha_j, \gamma)\Lambda_j^{(\gamma)}(k, t)|k|^{\alpha_j-1} \cdot \left[|k| - ik\beta_j\tan\left(\frac{\pi\alpha_j}{2}\right)\right], \tag{54}$$

---

[9]Extension to the case with N transport mechanisms is straightforward.

where we have defined the quantities:

$$\Lambda_j^{(\gamma)}(x,t) \equiv \lambda_j(x,t) \left[{}_0D_t^{1-\gamma} n \right](x,t), \quad j = 1,2. \tag{55}$$

The diffusive coefficients $C(\alpha,\gamma) \equiv A_\gamma \sigma^\alpha / \tau^\gamma$ are the same we already defined in Sec. 2 for the integrable cases.

The Fourier inverse of Eq. 54 can then be written explicitly by introducing again the Riemann-Liouville fractional differential operators (Eq. 79, Appendix B):

$$\frac{\partial n(x,t)}{\partial t} = -\sum_{j=1}^{2} \frac{C(\alpha_j,\gamma)}{2\cos\left(\frac{\pi\alpha_j}{2}\right)} \left[(1+\beta_j)\frac{\partial^{\alpha_j}}{\partial x^{\alpha_j}} + (1-\beta_j)\frac{\partial^{\alpha_j}}{\partial(-x)^{\alpha_j}} \right] \cdot \Lambda_j^{(\gamma)}(x,t), \tag{56}$$

which should be compared with Eq. 29, obtained in Sec. 2 for integrable CTRWs. We may comment that, in the extended CTRW case, an equation with a fractional time derivative in terms of the Caputo operator is not available due to the presence of the nonlinearity in the projectors, in contrast to what happened with Eq. 28 in the integrable case.

We finish this section by noting that the rather complicated Eq. 56 can again be written in a more familiar form when both step size pdfs are symmetric. Then, again using the Riesz operator (Eq. 82, Appendix B), we obtain:

$$\frac{\partial n(k,t)}{\partial t} = C(\alpha_1,\gamma)\frac{\partial^{\alpha_1}\Lambda_1^{(\gamma)}}{\partial|x|^{\alpha_1}} + C(\alpha_2,\gamma)\frac{\partial^{\alpha_2}\Lambda_2^{(\gamma)}}{\partial|x|^{\alpha_2}}. \tag{57}$$

4.3 Interpretation of FDEs Eqs. 56 and 57.

The main difference between Eq. 56 and Eq. 29 is that the quantity on which the spatial fractional derivative acts is the "projected density", $\Lambda_j^{(\gamma)}(x,t)$. Ignoring, for the time being, the fractional time-derivative by setting $\gamma = 1$, and considering only the derivative appearing in the term proportional to $(1+\beta_j)$, we have that (recall Eq. 78, Appendix B):

$$\frac{\partial^{\alpha_j}\Lambda_j^{(1)}}{\partial x^{\alpha_j}} \equiv \frac{1}{\Gamma(p-\alpha_j)}\frac{d^p}{dx^p}\int_{-\infty}^{x} \frac{\lambda_j(x',t)n(x',t)dx'}{(x-x')^{\alpha_j-p+1}}, \tag{58}$$

where p is the integer part of α_j. Again, this fractional derivative collects the contributions by all walkers that end up at x at time t *from $x' \leq x$* by means of the transport mechanism j. But, since the argument in the integral is the "projected" walker density $\lambda_j(x',t)n(x',t)$, only those locations x' for which $\lambda_j(x',t) \neq 0$ can contribute to the density of walkers at x. In the case in which the projector describes some instability threshold (as in Eq. 34), it follows that the first of the two contributions to Eq. 56 from the transport mechanism j simply corresponds to the fact that any change in walker density at point x and time t *can only come from points $x' \leq x$ at which, at that same time, that mechanism is active!* Analogously, the second contribution to the transport mechanism j (the term proportional to $(1-\beta_j)$) gives the contribution to the change in $n(x,t)$ from points with $x' \geq x$ where the mechanism j is active at time t.

Before discussing the non-Markovian ($\gamma < 1$) case, it is important to note that Eq. 56, in spite of being Markovian, *contains some sort of system memory*. It is the "memory-through-profile" mechanism associated to the transport mechanism j, which is contained in the projector $\lambda_j(x, t)$. As we already discussed in Sec 3, the previous history of the system, that has been carved into the system profile by past transport events, can in this way affect the future system evolution [83, 84].

Let's look now at the non-Markovian case with $\gamma < 1$, that allows modelling memory effects in a probabilistic manner associated with the "microscopic" waiting time pdf ψ. In this case, $\Lambda_j^\gamma(x', t)$ appearing in Eq. 56 is more complicated than just the "projected density" $\lambda_j(x, t)n(x, t)$ that we just discussed. Writing $\Lambda_j^\gamma(x', t)$ explicitly, it happens that (recall Eq. 78, Appendix B):

$$\Lambda_j^\gamma(x', t) = \frac{1}{\Gamma(m-1+\gamma)} \frac{d^m}{dt^m}\left[\int_0^t \frac{\lambda_j(x', t')n(x', t')}{(t-t')^{2-\gamma-m}} dt'\right],$$

with m the integer part of $1 - \gamma$. Note that $\Lambda_j^\gamma(x', t)$ may be now be nonzero even if $\lambda_j(x', t) = 0$ at time t. The reason is that this term now collects contributions from all past times $t' < t$ when $\lambda_j(x', t') \neq 0$. Again, this would mean that $\Lambda_j^\gamma(x', t)$ is determined by the values of the density of walkers *at all $t' < t$ when that particular mechanism was active!*

5 One Field Model: Some Numerical Results

In this section we will explore to what extent the simple one-field model given by the GME Eq. 46, together with the kernel Eq. 47, is capable of capturing the phenomenology that we aimed for at the beginning of this paper, and that is characteristic of magnetically confined plasmas. In particular, we are interested in showing that the simple model, for a meaningful choice of ψ and p, can exhibit power-degradation, canonical profiles, superdiffusive pulse propagation, and on-axis peaking for off-axis fueling, while being capable of accommodating transport channels without characteristic scales.

Inspired by the plasma phenomenology, we will assume that a critical threshold exists above which, a non-local, fast transport channel becomes active [78]. Below the threshold, a slower, local transport channel takes care of all the transport. The former is represented by a Lévy distribution with $\alpha = 1$, while the latter is represented by a Gaussian distribution[10]. For simplicity, we choose the threshold to be a critical gradient Z_c, and the step size pdf to be given by Eq. 33, with the projectors defined as in Eq. 34.

For the waiting time pdf we will use a **simple exponential law**, $\psi(t - t') = \tau_1^{-1}\exp(-(t - t')/\tau_1)$, so that only memory effects related to the memory-through-profile mechanism discussed above are considered.

Of course, the numerical model must "live" in a **finite system**. Therefore, **boundary conditions** must be imposed at the edges of the system. It is physically more appealing to

---

[10]The choice $\alpha = 1$ is motivated by the fact that this is one of the few cases in which an analytic expression for the Lévy distribution exists in real space (see Appendix A).

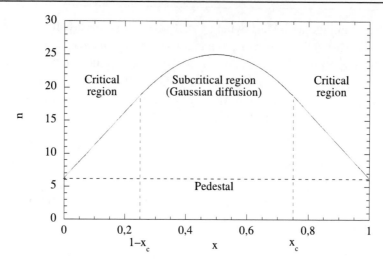

Figure 1: Sketch of the transport regions in the model for an intermediate fueling situation (From Ref. [78]. Copyright 2004 The American Institute of Physics).

do this with the GME instead of the corresponding fluid limit, since in the former case this reduces to truncating the limits of the integral in Eq. 46 to the system domain:

$$\frac{\partial n(x,t)}{\partial t} = S(x) - \frac{n(x,t)}{\tau_1} \tag{59}$$

$$+ \frac{1}{\tau_1} \int_0^1 dx' \left[\lambda_1(x',t) P_{(1,0,\sigma_1)}(x-x') + \lambda_2(x',t) P_{(2,0,\sigma_2)}(x-x') \right] \cdot n(x',t),$$

where it has been assumed that the system spans the interval $[0,1]$. Note that an **external fueling source** $S(x)$ must be added to compensate for the losses of particles through the system boundaries, required if the system is to reach some kind of steady state.

In the next sections, we will show some results obtained when numerically evolving Eq. 59 with the choices $\tau_1 = 1$, $Z_c = 50$, $\sigma_1 = 0.04$ and $\sigma_2 = 0.02$. First, using a uniform fueling source $S(x) = S_0$, we proceed to examine the steady state solution to look for signs of power degradation and canonical profiles (Sec. 5.1), non-diffusive scaling (Sec. 5.2) of the global confinement time and superdiffusive propagation (Sec. 5.3). Then, we use off-center local fueling and look for the formation of central density peaking (Sec. 5.4).

5.1 Scaling of the Global Confinement Time with the Strength of the External Fueling. Canonical Profiles

The one-field model has two limits as a function of the external fueling strength S_0: at weak fueling ($S_0 < S_c$), the anomalous transport channel is never activated since the gradient remains below critical everywhere; while at very strong fueling ($S_0 >> S_c$) the system behaves almost as a purely anomalous system, with a super-critical gradient ($|dn/dx| > Z_c$) almost everywhere. The **core** or **center of the system** (the region around $x = 1/2$) is an exception, because symmetry requires that the gradient is always zero in the center. Therefore, there always remains a central region where transport is diffusive (not anomalous) for any level of fueling, which shrinks in size as the fueling level increases.

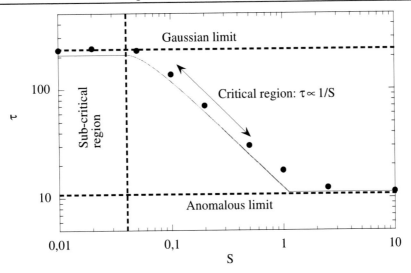

Figure 2: Confinement time as a function of the fueling rate S_0. The points are results from program runs, the continuous line is the theoretical relationship (which does not take account of the pedestal) derived in the text. For the parameters used, $S_c = 0.04$. For $S_0 < S_c$ the system is sub-critical and the confinement time does not depend on S_0. At low and intermediate values of S_0, the theoretical line agrees well with the experimental points, although there is a small difference, caused by the pedestal. Finally, at the highest values of S_0 criticality is lost in the periphery (the system is overdriven) and the experimental points tend asymptotically to the anomalous scaling limit (from Ref. [78]. Copyright 2004 The American Institute of Physics).

In an intermediate fueling situation, the system is divided into two main regions: a central region where transport is diffusive and a periphery where the gradient is critical (Fig. 1). In the central region, the slope of the profile can be computed, to good approximation, as:

$$\frac{dn}{dx} = -\left(\frac{S_0 \tau_1}{\sigma_2^2}\right)(x - 1/2). \tag{60}$$

In the critical region $Z = Z_c$ as a result of the *critical gradient being able to clamp the density gradient to the critical value*. The profile in this region thus becomes independent of the source strength or distribution: it is an example of a canonical profile.

The crossing-over point between the two regions is where $|dn/dx| = Z_c$:

$$\left|x_c - \frac{1}{2}\right| = \frac{\sigma_2^2}{S_0 \tau_1} Z_c \propto S_0^{-1}. \tag{61}$$

Finally, the critical fueling threshold that must be overcome for the anomalous channel to become active is given by:

$$S_c = \frac{2\sigma_2^2}{\tau_1} Z_c. \tag{62}$$

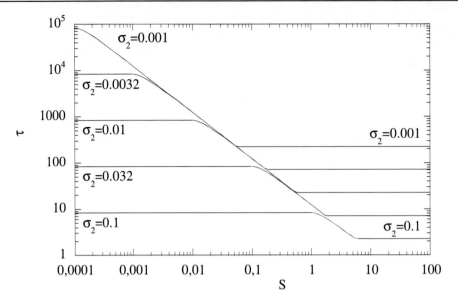

Figure 3: Scan of σ_2 and S_0 (from Ref. [78]. Copyright 2004 The American Institute of Physics).

Except for the pedestal contribution, the particle content of the system can easily be computed by integrating the curve sketched in Fig. 1:

$$
N_{partial} = \begin{cases} \dfrac{S\tau_D}{12\sigma_2^2} & (S < S_c) \\[4mm] \dfrac{1}{4}Z_c - \dfrac{\sigma_2^4}{3S^2\tau_D^2}Z_c^3 & (S \geq S_c) \end{cases}
\tag{63}
$$

The confinement time can simply be evaluated from $\tau = N_{tot}/S_0$. Taking into account that the result must tend to the anomalous limit (roughly given by $\tau \sim \tau_1/\sigma_1$ [53, 78]) for large S_0, we find:

$$
\tau_{partial} = \begin{cases} \dfrac{\tau_D}{12\sigma_2^2} & (S < S_c) \\[4mm] \max\left[\dfrac{1}{4S}Z_c - \dfrac{\sigma_2^4}{3S^3\tau_D^2}Z_c^3, \dfrac{c\tau_1}{\sigma_1}\right] & (S \geq S_c) \end{cases}
\tag{64}
$$

where c is a constant. This behavior is reflected by the scan of S_0 presented in Fig. 2. Clearly, power degradation is only observed in the intermediate fueling regime, while the confinement time becomes independent of S_0 when only one transport channel dominates. Interestingly, the observation of canonical profiles is also associated to this intermediate regime where the critical gradient is able of clamping the particle density profile. Thus, it is not strange to observe canonical profiles and power degradation simultaneously. Finally,

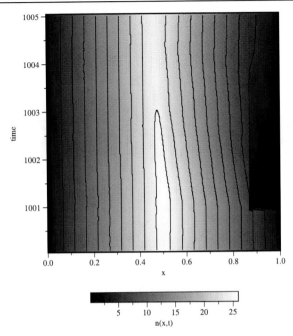

Figure 4: Graph of $n(x, t)$. Cold pulse induced at $t_{cp} = 1001$ (from Ref. [78]. Copyright 2004 The American Institute of Physics).

note that these results strengthen the relation to self-organized critical (SOC) models, as can be ascertained by comparing these confinement times with those obtained for an SOC sandpile model built on similar principles [85].

5.2   Scaling of the Confinement Time with System Size

The generic scaling behavior obtained in the previous section can also be used to understand the scaling of the confinement time with system size. Since we have normalized the system size to $L = 1$, a scaling of the system size is equivalent to a scaling of σ. The constant c appearing in Eq. (64) can be estimated numerically for the parameters used and we find $c = 0.45$. In Fig. 3, we have plotted $\tau_{partial}$ as a function of σ_2 and S_0 (with $\sigma_1 = 2\sigma_2$). At each choice of σ_2, the system behaves as in the previous section: at low S_0, the system is fully diffusive; at high S, the system is completely anomalous; and in-between, the system is critical. In the critical situation, the confinement time is determined almost exclusively by the fueling S and the critical gradient Z_c; the waiting time τ_1 and the step size σ_2 do not play any significant role – or, in other words, they do not act as the characteristic scales of the transport.

5.3   Rapid Transport Phenomena

To show that the one-field model also produces superdiffusive transport phenomena, we have induced an artificial "cold pulse" at the edge of the system by setting $n(x, t_{cp}) = 0$ for $x \geq 0.875$ at $t_{cp} = 1001$. The strength of the source in this case is $S_0 = 0.2$. Fig. 4

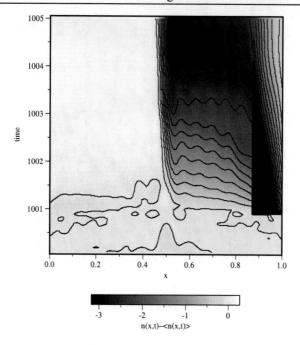

Figure 5: Graph of $n(x,t) - \langle n(x,t) \rangle$. Cold pulse induced at $t_{cp} = 1001$ (from Ref. [78]. Copyright 2004 The American Institute of Physics).

shows the time development of $n(x,t)$ afterwards. Clearly, the pulse creates an inward propagating "cold" front (i.e. with reduced density). Fig. 5 shows the same data, however after subtracting the steady state profile in order to stress the perturbation. The "cold" front reaches the center of the system ($x = 0.5$) in about $\Delta t = 0.1$. This number should be compared to the confinement time for these parameters: $\tau \simeq 70$. Thus, the front propagates nearly three orders of magnitude faster than might be expected from the global confinement time! Note that the propagation is "superdiffusive" in the sense that it does not slow down as it propagates, as would be expected for diffusive propagation.

5.4   Central Density Peaking with Off-Center External Fueling

To conclude, we will also examine the response of the one-field model to a nonuniform source. Inspired by the experimental observations in magnetically confined plasmas, we will consider local sources that are *symmetric with respect to the system center*, located at $x = 1/2$. In fact, we will use the following source function for off-center fueling:

$$S(x) = \frac{S_0}{2\sqrt{2\pi}w} \left(\exp\left[-\frac{(x-0.3)^2}{2w^2}\right] + \exp\left[-\frac{(x-0.7)^2}{2w^2}\right] \right), \qquad (65)$$

where we have chosen $S_0 = 0.2$ and $w = 0.025$. For reference, we show the response of the system in the absence of critical gradients and Lévy distributions, and with $\sigma_2 = 0.02$ and $\tau_1 = 1$. The resulting profile is shown in Fig. 6, exhibiting the usual diffusive behavior leading to a flat profile inside the source location.

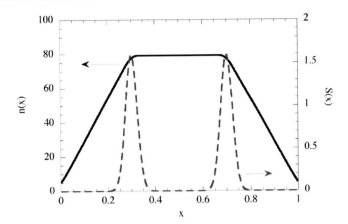

Figure 6: Density profile $n(x)$ of the one-field model forced with off-axis fueling: purely diffusive case (from Ref. [86]. Copyright 2004 The American Institute of Physics).

Next, we activate the critical gradient and the Lévy distributions. The steady state solution is shown in Fig. 7. Note the stark difference of the response when compared to the diffusive situation with identical source profile (Fig. 6). The most relevant feature is that *the profile peaks in the central region*, where $S(x) = 0$! The latter is remarkable - for *how can a gradient be maintained in this region, without a central source?* Transport around and outside of the source region is self-organizing around the critical gradient. The self-organization requires that the rapid (supercritical) transport channel, here associated with non-Gaussian Lévy distributions, is activated at least part of the time in the peripheral regions. These Lévy distributions are characterized by the fact that a small but significant part of the particles take "long steps" and end up in the central region. Thus, the central region fills up with particles originating in the critical region (note, however, that this mechanism does not rely critically on the use of Lévy distributions, while it does rely on the existence of a fast and slow transport channel [87]). When the same experiment is repeated at a lower value of S_0, the result shown in Fig. 8 is obtained. In this case, off-axis fueling indeed leads to a mostly flat (or only slightly curved) profile in the core region.

To study the transition between the situation of Fig. 7 and Fig. 8, we have computed several more cases, varying only the total fueling rate S_0. Fig. 9 shows the **density peaking factor**, defined as the quotient between the core density and the average density, as a function of the source strength. The peaking factor is small for the smallest fueling strengths (because of the flat central profile), but increases rapidly with S_0 until it saturates at a value of about 1.85. The same figure also shows the size of the central sub-critical region. Recall that the location of the crossing point between the core and the critical region, x_c, is given by Eq. 61. Thus, $|x_c - 0.5|$ is the half-width of the central sub-critical region. From the figure, it is clear that when the sub-critical region encompasses the fueling location (at $|x - 0.5| = 0.2$), the profile peaking is low since fueling is done in a sub-critical region, while profile peaking becomes important as soon as the fueling takes place in a region that is locally critical.

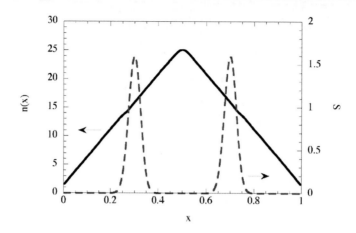

Figure 7: Density profile $n(x)$ with strong off-center fueling: Generalized master equation ($S_0 = 0.2$) (from Ref. [86]. Copyright 2004 The American Institute of Physics).

This profile peaking is however not only sensitive to changes in the source strength S_0. It also appears when *the critical gradient Z_c is changed* and when the *ratio σ_2/σ_1, that measures the relative strength of the slow and fast transport channels, changes*.

To see this, we first note that Eq. 59 is linear in S, which means that a multiplication of S by a factor C leads to a multiplication of n by the same factor C, *provided* the dependency on n inside the functional specification of p is adjusted accordingly. In other words, when S and Z_c are both multiplied by C, the corresponding steady state profile is $C \cdot n(x)$, and the profile has the same shape. Thus, the above experiment, in which S was lowered by a factor of 4 while Z_c was held constant, is equivalent to an experiment in which S is kept constant while Z_c is raised by a factor of 4. The experimentally observed difference in steady state profiles with off-axis fueling can thus either be attributed to a difference in fueling levels (as in the example given), or to a different value of the critical gradient.

Regarding the effect of changing the ratio of the scales (σ) of the diffusive (Gaussian) to the anomalous (Lévy) step size distributions, we have increased the diffusive transport channel by increasing σ_2. Fig. 10 shows the system response as a function of σ_2, with off-axis fueling as above ($S_0 = 0.2$). Since $D = \sigma_2^2/\tau_1$ for the diffusive (Gaussian) transport channel, this is equivalent to scanning the diffusion coefficient D while keeping the anomalous transport parameters ($\sigma_1 = 0.04$ and $Z_c = 50$) constant.

As σ_2 increases, Gaussian diffusion starts to dominate, lowering the gradients so that the system becomes more sub-critical. This is reflected by the gradients at $\sigma_2 = 0.04$, which are mostly well below the critical value (compare with the profile at $\sigma_2 = 0.02$, which exhibits critical values of the gradient in most of the system), except immediately outside the fueling position, where the drive is sufficient to maintain a locally supercritical gradient. The anomalous character of the system can be seen from the central bulge, which should be absent in a purely diffusive system (since the source S is zero in the core region). The

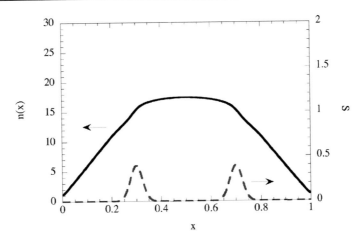

Figure 8: Density profile $n(x)$ with weak off-center fueling: Generalized master equation $(S_0 = 0.05)$ (from Ref. [86]. Copyright 2004 The American Institute of Physics).

bulge is due, of course, to particles originating in the supercritical zone taking long steps. Finally, at $\sigma_2 = 0.06$ nearly the whole system appears to be sub-critical or diffusive, and the response to off-axis fueling is rather similar to that expected from a Gaussian diffusive model.

6 Conclusions

In this paper, we have reviewed a very recent extension of the standard CTRW/GME that might provide a very rich framework capable of capturing some of the complex phenomenology observed in plasma turbulent transport studies in tokamaks and stellarators. Although it is currently formulated for only a single field (i.e., the particle density), we have shown that its formal structure is capable of handling non-local transport properly and, by considering non-integrable CTRW/GMEs, phenomena such as power degradation, canonical profiles, superdiffusive propagation and profile peaking in the presence of off-axis heating may be produced in a natural way. For this reason, we think that this approach may provide the first step towards the development of more complex models (capable of dealing also with temperature transport) in the near future.

As it is, the formalism may be applied either through the full GME (Eq. 46) or by taking the fluid limit (Eq. 57). From an aesthetic point of view, the latter takes us closer to the classical diffusive equations, but at the price of having to deal with FDEs. An argument against the use of FDEs might be that boundary conditions can be imposed in a much more intuitive way for GMEs. However, the FDE approach has also some advantages of its own. The more interesting of these, it seems to us, is that it *allows combining transport mechanisms that do not share the same temporal dynamics* [81].

This is a somewhat subtle, but important issue. Note that the derivation of the extended CTRW/GME presented here is based on an important assumption that is not always justified

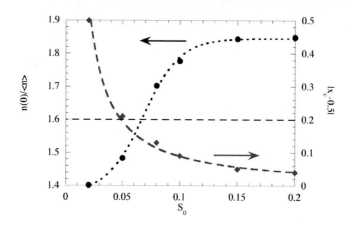

Figure 9: Density peaking factor (left axis) and half-width of the central subcritical region (right axis) vs. S_0 for off-axis fueling (from Ref. [86]. Copyright 2004 The American Institute of Physics).

in practice: namely, that all transport mechanisms share the same temporal dynamics and characteristic scales, in the sense that the same waiting time pdf ψ is used at all times, independently of which transport channel is active. However, and coming back to the example of a magnetically confined plasma, it is well known that the two transport channels that set the dynamics of particle and energy transport in these plasmas – collisional diffusion and turbulence– have different associated time scales. In particular, both timescales can change is a very different manner when external parameters such as the plasma temperature or the strength of the magnetic field are varied [5]. The FDE approach provides a way to handle this situation, because the fluid limit of each individual transport mechanism is invariant under the rescaling:

$$\{\gamma, \tau, \alpha, \beta, \sigma\} \rightarrow \{\gamma, \tau', \alpha, \beta, \sigma'\}. \tag{66}$$

provided:

$$[\sigma'/\sigma]^\alpha = [\tau'/\tau]^\gamma. \tag{67}$$

Therefore, as long as the *essential* temporal and spatial dynamics of each individual CTRW remain unchanged (by essential dynamics, we mean the values of α, γ and β), it is always possible to rescale the temporal and spatial scale parameters of all CTRWs (which are associated to the temporal and spatial characteristic scales of each transport mechanism) so that their rescaled temporal scale parameters are all the same. On the other hand, the value of γ cannot be rescaled in this fashion. For this reason, consideration of several transport mechanisms with different *essential* temporal dynamics is not possible in this framework, not even in the fluid limit.

Finally, we would like to conclude this paper by making some comments on the relationship between macroscopic transport based on Lévy statistics and the microscopic motion of the particles that make up the plasma. The way we envision this relationship is

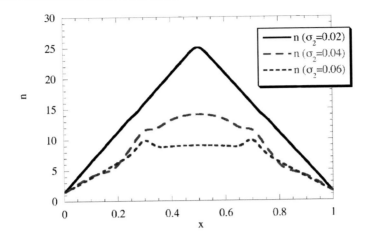

Figure 10: Density profile $n(x)$ with off-center fueling for varying diffusive transport parameter σ_2 (from Ref. [86]. Copyright 2004 The American Institute of Physics).

the following: Lévy distributions appear when describing transport at a *mesoscale* range, intermediate between the microscopic motion of the particles and the fluid limit, when the transport events (avalanches) can be substituted by a probabilistic description. This probabilistic description is provided by the Lévy distributions, that capture the lack of characteristic scales of the avalanches. Naturally, it would be much more desirable to obtain such an intermediate regime by a proper averaging of the relevant kinetic equations. However, this is a remarkably difficult problem as has been pointed out elsewhere [58].

Acknowledgments

Many enlightening discussions (both direct and electronic) with M. Varela, D.E. Newman, P.W. Terry, R. Balescu, R.A. Woodard, G.M. Zaslavsky, C. Hidalgo, L. García, J.A. Mier, J.R. Martín-Solís and D. del-Castillo-Negrete are acknowledged. Research supported in part by Spanish DGES Projects No. FTN2003-04587 and FTN2003-08337-C04-01. Research carried out in part at Oak Ridge National Laboratory, managed by UT-Battelle, LLC, for U.S. DOE under contract number DE-AC05-00OR22725.

A   Lévy Distributions

The Lévy-Gnedenko family of pdfs comprises all the possible limit distributions that are strictly stable with respect to the *sum of N independent and identically distributed (i.i.d.) random variables* [43, 44]. The family is defined in terms of three parameters, and its members are denoted by $P_{\alpha,\beta,\sigma}(y)$. They can be defined in terms of their Fourier transform or characteristic function as ($0 < \alpha \leq 2, |\beta| \leq 1$) [44]:

$$P_{\alpha,\beta,\sigma}(k) =$$
$$= \exp\left[-\sigma^\alpha |k|^\alpha \left(1 - i\beta \; \mathrm{sgn}(k) \tan\left(\frac{\pi\alpha}{2}\right)\right)\right]. \tag{68}$$

The three labels define the properties of each distribution. First, β measures the *asymmetry* of the distribution. This comes from the fact that:

$$P_{\alpha,\beta,\sigma}(y) = P_{\alpha,-\beta,\sigma}(-y). \tag{69}$$

β can vary between $-1 \leq \beta \leq 1$ for all $\alpha \neq 1, 2$, while for the latter values of α only $\beta = 0$ is possible. Secondly, α gives the asymptotic behaviour of the distribution at large y. Thus, for $0 < \alpha < 2$ all Lévy distributions exhibit heavy tails. Indeed, for $\alpha \neq 1$, one has:

$$P_{\alpha,\beta,\sigma}(y) \sim \begin{cases} C_\alpha \left(\frac{1-\beta}{2}\right) \sigma^\alpha |y|^{-(1+\alpha)}, & y \to -\infty \\ \\ C_\alpha \left(\frac{1+\beta}{2}\right) \sigma^\alpha |y|^{-(1+\alpha)}, & y \to +\infty \end{cases} \tag{70}$$

where the constant is given by:

$$C_\alpha = \frac{(\alpha-1)\alpha}{\Gamma(2-\alpha)\cos(\pi\alpha/2)}, \tag{71}$$

$\Gamma(x)$ is Euler gamma function. In the special case $\alpha = 1$, the PDF decays as $P_{1,0,\sigma}(y) \sim (\sigma/\pi)|y|^{-2}$. Finally, σ is called a *scale parameter* because:

$$P_{\alpha,\beta,\sigma}(ay) = P_{\alpha,\,\mathrm{sgn}(a)\beta,|a|\sigma}(y). \tag{72}$$

Extremal Lévy Distributions

A Lévy distribution is called *extremal* when its skewness value is maximum: $\beta = \pm 1$ for $\alpha \neq 1, 2$. It is important to note that, according to the previous equations, the power law decay is only observed in one tail for such extremal distributions ($\beta = \pm 1$), the other tail decaying exponentially. In the case of $1 < \alpha < 2$, $\beta = +1$ implies that the exponential tail exists for $y \to \infty$, while $\beta = -1$ has an right exponential tail for $y \to -\infty$. For $0 < \alpha < 1$ the extremal distributions are *one-sided* [70]: they are defined only for $y > 0$ if $\beta = +1$ and for $y < 0$ if $\beta = -1$. In that case, the exponential tail is found in the limit $y \to 0+$ for $\beta = -1$, and for $y \to 0-$ for $\beta = 1$. Their Laplace transform is given by:

$$P_{\alpha,1,\sigma}(s) = \exp\left[-\frac{\sigma^\alpha}{\cos(\pi\alpha/2)}s^\alpha\right]. \tag{73}$$

Moments of Lévy Distributions

Another important property of the Lévy distributions is that all moments higher than α are infinite. That is, the momenta of $P_{\alpha,\beta,\sigma}$ verify:

$$<|x|^p> = \begin{cases} \infty, & p \geq \alpha \\ \\ [c_{\alpha,\beta}(p)]^p \sigma^p, & p < \alpha \end{cases} \tag{74}$$

where the coefficient is not relevant for our discussion (it is given in Ref. [44]). Thus, only the Gaussian distribution ($\alpha = 2$) has finite variance. Furthermore, all distributions with $\alpha \leq 1$ also have infinite first moments.

Explicit Expressions of Lévy Distributions

There are only three Lévy distributions for which an analytical expression exists [44]:
(i) The *Cauchy distribution*. Its real space representation is:

$$P_{1,0,\sigma}(y) = \frac{\sigma}{\pi(y^2 + \sigma^2)}, \tag{75}$$

(ii) the *Gauss distribution*,

$$P_{2,0,\sigma}(y) = \frac{1}{2\sigma\sqrt{\pi}} e^{-y^2/4\sigma^2}, \tag{76}$$

(note that the relation of σ with the usual width w of the Gaussian is thus $2\sigma^2 = w^2$) and
(iii) the Lévy distribution

$$P_{1/2,1,\sigma}(y) = \left(\frac{\sigma}{2\pi}\right)^{1/2} \frac{1}{y^{3/2}} e^{-\sigma/2y}. \tag{77}$$

B Fractional Differential Operators

The *Riemann-Liouville fractional derivative operators* can be defined explicitly by means of the integral operators [64, 62]:

$$_aD^\alpha_x f(x) \equiv \frac{1}{\Gamma(p-\alpha)} \frac{d^p}{dx^p} \left[\int_a^x \frac{f(x')dx'}{(x-x')^{\alpha-p+1}} \right], \tag{78}$$

$$^bD^\alpha_x f(x) \equiv \frac{-1}{\Gamma(p-\alpha)} \frac{d^p}{d(-x)^p} \left[\int_x^b \frac{f(x')dx'}{(x'-x)^{\alpha-p+1}} \right].$$

In these expressions, $\Gamma(x)$ is the usual Euler Gamma function, and p represents the integer part of $\alpha + 1$. a [or b] is called the start [end] point of the operator.

In the cases in which the start point a or the end point b extend all the way to infinity, we will use the notation:

$$\frac{d^\alpha f}{dx^\alpha} \equiv {}_{-\infty}D^\alpha_x f(x);$$

$$\frac{d^\alpha f}{d(-x)^\alpha} \equiv {}^{+\infty}D^\alpha_x f(x) \tag{79}$$

These operators are particularly interesting since their Fourier transforms satisfy [64, 62]:

$$F\left[\frac{d^\alpha f}{dx^\alpha}\right] = (-iq)^\alpha f(q), \tag{80}$$

$$F\left[\frac{d^\alpha f}{d(-x)^\alpha}\right] = (iq)^\alpha f(q). \tag{81}$$

Indeed, Eqs. 80– 81 may also be considered defining equations for these operators.

Another useful fractional operator is the so-called *Riesz fractional derivative operator* [64, 62]. It is defined symmetrically:

$$\frac{d^\alpha}{d|x|^\alpha} \equiv -\frac{1}{2\cos(\pi\alpha/2)}\left[\frac{d^\alpha}{dx^\alpha} + \frac{d^\alpha}{d(-x)^\alpha}\right]. \tag{82}$$

Thus, the Fourier transform of the Riesz operator satisfies:

$$F\left[\frac{d^\alpha f}{d|x|^\alpha}\right] = -|q|^\alpha f(q), \tag{83}$$

which follows from Eqs. 80–81 due to the complex identity:

$$(-iq)^\alpha + (iq)^\alpha = 2|q|^\alpha \cos\left(\frac{\pi\alpha}{2}\right), \tag{84}$$

where $i = \sqrt{-1}$, the usual imaginary unit.

Finally, the *Caputo fractional derivative operator* is defined as [67]:

$$\frac{d_c^\gamma f}{d_c x^\gamma}(x) \equiv \frac{1}{\Gamma(\gamma - p)}\int_0^x \frac{d^p f}{dx^p}(x')\frac{d\tau}{(x - x')^{\gamma+1-p}}, \tag{85}$$

where p is the integer part of γ. The Caputo fractional derivative is usually associated to derivatives in time. The need for defining a different fractional derivative when time is involved (instead of using the Riemann-Liouville operator with starting point at $t = 0$) is related to the fact that the Laplace transform of the Caputo derivative satisfies [62, 64]:

$$\mathsf{L}\left[\frac{d_c^\gamma f}{d_c t^\gamma}(t)\right] = s^\gamma f(s) - \sum_{k=0}^{p-1} s^{\gamma-k-1}\cdot\frac{d^k f}{dt^k}(0), \tag{86}$$

which depends only on the initial values of $f(t)$ and its integer derivatives. In contrast, the Laplace transform of $_0D^\gamma_t f(t)$ depends on $f(t)$ and the initial values of fractional derivatives of lower order than γ, which do not have a clear physical meaning in the case of practical applications [62, 64]. The relation between Riemann-Liouville and Caputo derivatives is given by [64]:

$$_0D^\gamma_t f(t) = \frac{d_c^\gamma f}{d_c t^\gamma} + \frac{t^{-\gamma}f(0)}{\Gamma(1 - \gamma)}. \tag{87}$$

References

[1] J. Wesson, *Tokamaks*, Oxford Univ. Press, 3rd. Edition, Oxford (1998)

[2] M. Wakatani, *Stellarator and Heliotron Devices*, Oxford Univ. Press, Oxford (1998)

[3] G.H. Neilson, E.A. Lazarus *et al*, *Nucl. Fusion* **23**, 285 (1983)

[4] R.J. Goldston, *Plasma Phys. Control. Fusion* **26**, 87 (1984)

[5] ITER Physics Experts Group on Confinement and Transport Modelling and Database, *Nucl. Fusion* **39**, 2176 (1999)

[6] F.L Hinton and R.D. Hazeltine, *Rev. Mod. Phys.* **48**, 239 (1976)

[7] S.P. Hirshman and D.J. Sigmar, *Nucl. Fusion* **21**, 1079 (1981)

[8] S. Chapman and T. G. Cowling, *The Mathematical Theory of Non-Uniform Gases*, Cambridge University Press, (1970)

[9] A.J. Wooton, B.A. Carreras et al, *Phys. Fluids B* **2**, 2879 (1990)

[10] J. Weiland and A. Hirose, *Nucl. Fusion* **32**, 151 (1992)

[11] G.W. Hammet and F.W. Perkins, *Phys. Rev. Lett.* **64**, 3019 (1990)

[12] R.E. Waltz, R.R. Dominguez and G.W. Hammet, *Phys. Fluids B* **4**, 3138 (1992)

[13] W.W. Lee, *J. Comput. Phys.* **72**, 243 (1987)

[14] R.D. Sydora, *Phys. Fluids B* **2**, 1455 (1990)

[15] A.M. Dimits, T.J. Williams, J.A. Beyers and B.I. Cohen, *Phys. Rev. Lett.* **77**, 71 (1996)

[16] M. Kotschenreuter, W. Dorland et al, *Phys. Plasmas* **4**, 2381 (1995)

[17] J.E. Kinsey and G. Batemann, *Phys. Plasmas* **5**, 3344 (1996)

[18] R.E. Waltz, G.M. Staebler et al, *Phys. Plasmas* **4**, 2482 (1997)

[19] B. Coppi, *Comments Plasma Phys. Controlled Fusion* **5**, 261 (1980)

[20] C.C. Petty and T.C. Luce, *Nucl. Fusion* **34**, 121 (1994)

[21] F. Ryter, G. Tardini et al, *Nucl. Fusion* **43**, 1396 (2003)

[22] K.W. Gentle, R.V. Bravanec et al, *Phys. Plasmas* **2**, 2292 (1995)

[23] J.D. Callen and G.L Jahns, *Phys. Rev. Lett.* **38**, 491 (1977)

[24] N.J. Lopes-Cardozo, *Plasma Phys. Contr. Fusion* **37**, 799 (1995)

[25] J.G. Cordey, D.G. Muir *et al.*, *Nucl. Fusion* **35**, 101 (1995)

[26] E.D. Fredrickson, K. McGuire *et al.*, *Phys. Rev. Lett.* **65**, 2869 (1990)

[27] M.W. Kissick, J.D. Callen and E. Fredrickson, *Nucl. Fusion* **38**, 821 (1998)

[28] B.Ph. van Milligen, E. de la Luna *et al*, *Nucl. Fusion* **42** (2002) 787

[29] H. Biglari and P.H. Diamond, *Phys. Fluids B* **3**, 1797 (1991)

[30] D.R. Baker, C.M. Greenfield *et al*, *Phys. Plasmas* **8**, 4128 (2001)

[31] F. Jenko, W. Dorland, M. Kotschenreuther and B.N. Rogers, *Phys. Plasmas* **7**, 1904 (2000)

[32] W. Dorland, F. Jenko, M. Kotschenreuther and B.N. Rogers, *Phys. Rev. Lett.* **85**, 5579 (2000)

[33] P. Bak, C. Tang and K. Wiesenfeld, *Phys. Rev. Lett.* **59**, 381 (1987).

[34] P.H. Diamond and T.S. Hahm, *Phys. Plasmas* **2**, 3640 (1995)

[35] D.E. Newman, B.A. Carreras, P.H. Diamond and T.S. Hahm, *Phys. Plasmas* **3**, 1858 (1996)

[36] B.A. Carreras, D.E. Newman, V.E. Lynch and P.H. Diamond, *Phys. Plasmas* **3**, 2903 (1996)

[37] R. Sánchez, D.E. Newman and B.A. Carreras, *Nucl. Fusion* **41**, 247 (2001)

[38] B.A. Carreras, B.Ph. van Milligen *et al*, *Phys. Rev. Lett.* **80**, 4038 (1998)

[39] M.A. Pedrosa, C. Hidalgo *et al*, *Phys. Rev. Lett.* **82**, 3621 (1999)

[40] B.A. Carreras, B.Ph. van Milligen *et al*, *Phys. Rev. Lett.* **83**, 3653 (1999)

[41] P.A. Politzer, *Phys. Rev. Lett.* **84**, 1192 (2000)

[42] R. Sánchez, B. Ph. van Milligen, D.E. Newman and B.A. Carreras, *Phys. Rev. Lett.* **90**, 185005 (2003)

[43] B.V. Gnedenko and A.N. Kolmogorov, *Limit distributions for sums of independent random variables*, Addison Wesley, Reading, MA (1954)

[44] G. Samorodnitsky and M.S. Taqqu, *Stable non-Gaussian processes*, Chapman & Hall, New York (1994)

[45] E.W. Montroll and G. Weiss, *J. Math. Phys.* **6**, 167 (1965)

[46] E.W. Montroll and M.F. Shlesinger, in *Studies in Statistical Mechanics*, Ed. J.L. Lebowitz and E.W. Montroll (North-Holland, Amsterdam, 1984), Vol. 11, p.5

[47] H. Scher and M. Lax, *Phys. Rev. B* **7**, 4491 (1972)

[48] V.M. Krenke, E.W. Montroll and M.F. Shlesinger, *J. Stat. Phys.* **9**, 45 (1973)

[49] J. Klafter and R. Silbey, *Phys. Rev. Lett.* **44**, 55 (1980)

[50] E. Scalas, R. Gorenflo and F. Mainardi, *Phys. Rev. E* **69**, 011107 (2004)

[51] W. Feller, *An introduction to probability theory and its applications*, John Wiley & Sons, New York 1966.

[52] R. Metzler and J. Klafter, *Physics Reports* **339**, 1 (2000)

[53] G.M. Zaslavsky, *Physics Reports* **371**, 461 (2002)

[54] V.M. Krenke and R.S. Knox, *Phys. Rev. B* **9**, 5279 (1974)

[55] W. Shugard and H. Reiss, *J. Chem. Phys.* **65**, 2827 (1976)

[56] G.H. Weiss, *Aspects and applications of random walks*, North Holland, Amsterdam (1994)

[57] R. Balescu, *Phys. Rev. E* **51**, 4807 (1995)

[58] R. Balescu, *Aspects of Anomalous Transport in Plasmas*, Institute of Physics, Bristol (2005).

[59] B. Berkowitz, J. Klafter, R. Metzler and H. Scher, *Wat. Resourc. Res.* **38** 1191 (2002)

[60] B.A. Carreras, V.E. Lynch, D.E. Newman and G.M. Zaslavsky, *Phys. Rev. E* **60**, 4770 (1999)

[61] A.I. Saichev and G.M. Zaslavsky, *Chaos* **7**, 753 (1997)

[62] K. Oldham and J. Spanier, *The fractional calculus*, Academic Press, New York 1974.

[63] A. Carpinteri and F. Mainardi, *Fractals and fractional calculus in continuum mechanichs*, Springer-Verlag, New York (1997)

[64] I. Podlubny, *Fractional differential equations*, Academic Press, New York 1998.

[65] J.T. Edwards, N.J. Ford and A.C. Simpson, *Journal of Computational and Applied Mathematics* **148**, 401 (2002)

[66] V.E. Lynch, B.A. Carreras, D.del-Castillo-Negrete, K.M. Ferreira-Mejías and H.R. Hicks, *J. Comput. Phys.* **192**, 406 (2003)

[67] M. Caputo, *Geophys. J. Royal Astron. Soc.* **13**, 529 (1967)

[68] R. Hilfer, *J. Phys. Chem. B* **104**, 3914 (2000)

[69] F. Mainardi, M. Raberto, R. Gorenflo and E. Scalas, *Physica A* **287**, 468 (2000)

[70] R. Gorenflo, in *Fractals and fractal calculus in continuum mechanics*, Ed. A. Carpinteri and F. Mainardi (Springer-Verlag, Wien, 1997).

[71] X. Garbet, P. Mantica et al, *Plasma Phys. Contr. Fus.* **46**, 1351 (2004)

[72] J.P. Freidberg, *Ideal magnetohydrodynamics*, Plenum Press, New York (1987)

[73] X. Garbet and R. Waltz, *Phys. Plasmas* **5**, 2836 (1998)

[74] Y. Sarazin and P. Ghendrih, *Phys. Plasmas* **5**, 4214 (1998)

[75] G. M. Zaslavsky, M. Edelman *et al*, *Phys. Plasmas* **7**, 3691 (2000)

[76] B.A. Carreras, V.E. Lynch and G.M. Zaslavsky, *Phys. Plasmas* **8**, 5096 (2001)

[77] D. del-Castillo-Negrete, B.A. Carreras and V.E. Lynch, *Phys. Plasmas* **11**, 3854 (2004)

[78] B.Ph. van Milligen, R. Sánchez and B.A. Carreras, *Phys. Plasmas* **11**, 2272 (2004)

[79] R. Sánchez, B.Ph. van Milligen and B.A. Carreras, *Phys. Plasmas* **12**, 056105 (2005)

[80] E. Scalas, R. Gorenflo and F. Mainardi, *Physica A* **284**, 376 (2000)

[81] R. Sánchez, B.A. Carreras and B.Ph. van Milligen, *Phys. Rev. E* **71**, 011111 (2005)

[82] B.Ph. van Milligen, B.A. Carreras and R. Sánchez, *Plasma Phys. Control. Fusion* **47** (2005) B743

[83] P. Bantay and I.M. Janosi, *Phys. Rev. Lett.* **68**, 2058 (1992)

[84] A. Diaz-Guilera, *Europhys. Lett.* **26**, 177 (1994).

[85] R. Sánchez, D.E. Newman, B.A. Carreras, R. Woodard, W. Ferenbaugh and R. Hicks, *Nucl. Fusion* **43**, 1031 (2003)

[86] B.Ph. van Milligen, B.A. Carreras and R. Sánchez, *Phys. Plasmas* **11**, 3787 (2004)

[87] J. Gavnholt, J. Juul Rasmussen, O. E. Garcia, V. Naulin, and A. H. Nielsen, *Phys. Plasmas* **12**, 084501 (2005)

In: Advances in Plasma Physics Research, Volume 7
Editor: Francois Gerard

ISBN: 978-1-61122-983-7
© 2011 Nova Science Publishers, Inc.

Chapter 4

EXPERIMENTAL STUDY ON TANDEM MIRROR EDGE PLASMAS AND THEORETICAL STUDY ON PLASMAS IN DIVERTOR ENVIRONMENT

Md. Khairul Islam[1] and Yousuke Nakashima

Plasma Research Center, University of Tsukuba, Tsukuba, Ibaraki 305-8577, Japan
[1]Bangladesh Atomic Energy Commission, P.O. Box No.:158, Dhaka 1000, Bangladesh

ABSTRACT

A rigorous study on edge plasma in the anchor cell of the GAMMA 10 tandem mirror is carried out for the first time using Langmuir probes, calorimeters, conducting plates (anchor plate (AP)), and H_α detectors. Computational studies on the magnetic filed configuration (MFC) of the GAMMA 10 and neutral transport phenomena are carried out to understand the experimental results. Probe current asymmetry is found in the minor axis direction of the elliptic flux tube at the non-axisymmetric MFC region. Measurements by calorimeters and APs support this plasma asymmetry. Some impurity deposited areas are observed on the APs and the pattern of these areas is consistent with the plasma asymmetry. Formation of comparatively high density, but low temperature plasma in the anchor cells is observed, especially during axial confinement. Significant effect of the floated AP on the GAMMA 10 plasma parameters is observed. Using three dimensional Monte-Carlo code, a modeling of neutral transport is successfully performed in the GAMMA 10 anchor cell with non-axisymmetric MFC, and the behavior of neutrals in this region is investigated on the basis of the experimental data of H_α intensities. The computational result well predicts the experimental result. Ambient neutral pressure control experiment with modulating gas fueling rate is performed and significant improvement of anchor plasma parameters is obtained. Possible explanations of these observations are given in detailed and adverse effects of no-axisymmetric MFC on GAMMA 10 plasma parameters are also pointed out. Finally, possible way to optimize the non-axisymmetric MFC of the anchor cell is discussed in part A of this chapter. Part B is concerned with the results of theoretical investigation on the properties and excitation of low frequency electrostatic dust modes, e.g., dust-acoustic (DA) and dust-lower-hybrid (DLH) waves, in a divertor plasma environment using the fluid model. In this study, dust grain charge is considered as a dynamical variable in streaming magnetized dusty plasmas with a back ground of neutral atoms/molecules. Dust charge fluctuation, collisional, and streaming effects on DA and DLH modes are discussed. Charging of dust grains is also addressed.

1. Introduction

Edge plasma in magnetic fusion devices has important role in improving core plasma parameters. Atomic process, collision, dust grain formation, radiation loss, and a rich variety of possible plasma wall interactions (PWIs) are present in edge plasma. Relation between core plasma and edge plasma is not fully clarified, but it is clear that edge conditions significantly control the L-H transition in tokamak plasma [1-3]. Particle confinements, i.e., plasma and impurity densities are often strongly controlled by edge conditions, which are understood from the experiments in early days [4-6]. For steady state operation of next step fusion devices, conditions of edge plasma may become crucial issues for both energy and particle confinements of plasma. Core plasma behavior can also be understood by studying edge plasma properties. Under these circumstances, edge plasma are extensively investigated – experimentally and theoretically – both in closed and open fusion devices to optimize edge conditions in order to improve core plasma parameters [7-10].

There are many approaches to magnetic confinement for studying plasma confinement properties and heating mechanisms. Ability to avoid or, minimize the harmful effects of both magnetohydrodynamical (MHD) and micro-instabilities in plasma confinement is the basic criteria to select these approaches. These approaches can be divided into two categories: 1) Open system: In which the magnetic lines of forces leave confinement region, such as magnetic mirror. 2) Closed system: In which the magnetic lines of forces remain within confinement region, such as tokamak. Other prominent differences between open and closed systems are: 1) To minimize radial transport and to increase plasma β in tokamaks and similar toroidal devices are the important issues, whereas, linear mirror systems are inherently high β devices in which improvement of axial confinement is the main problem. 2) Engineering design of linear systems is simple compared to toroidal systems. 3) Open system can be operated in steady state. Successful axial confinement can make possible to develop smaller power plants based on linear mirror systems as compared to the minimum-size economical toroidal system. These advantages of linear geometry are the driving motivations for mirror fusion research. Tandem mirror is the improved configuration based on basic mirror concept [11,12]. In the GAMMA 10 tandem mirror, ion temperature of $10keV$ [13] and axial confinement potential of $2.2keV$ [14] are attained in the hot ion mode of operation. However, there are many problems yet to be solved before achieving a fusion reactor with the mirror system. Understanding the phenomena related with the problem for increasing plasma density in GAMMA 10 at the temperature of $10keV$ is very much important. In the competition of clarifying the physics of low β plasma confinement, close system is in successful state comparing to the open system. Though, TARA [15], Phaedrus [16], TMX-U [17], and MFTF-B [18,19] tandem mirrors have been shut down, big projects like GAMMA 10 [13,14,20], AMBAL-M [21,22], GDT [23], HANBIT [24], etc. are currently operated to understand the physical phenomena of the tandem mirror plasma. GAMMA 10 is the largest tandem mirror in the world.

Various kinds of plasma transport process lead strong PWI and the properties of edge plasma are strongly affected by PWI. Due to the difference in the geometry between open and closed systems, the mechanism of occurring PWI and the control of PWI are different between these two systems [8,10,25,26]. Using a magnetic divertor in a tokamak, many problems associated with PWI may be avoided or reduced. Particularly in the JET [27] and TEXTOR [28] tokamaks, edge plasma behavior and transport processes are investigated intensively. In a tandem mirror, magnetic lines of forces are ended in the metallic plates at end regions and axial confinement potential is almost absent in the edge region. Therefore, ions or electrons from the core plasma that transport/diffuse in the edge region flow along the field lines up to the end regions, where they are neutralized and removed by vacuum pumping [29]. This natural divertor action reduces PWI in open systems and moreover, shields the core plasma against the influx of recycling particles. Local PWI due to strong transport processes and the consequent influx of recycling particles can also be controlled using limiter. PWIs in the central cell, plug/barrier cells, and end cells of the TMX and TMX-U tandem mirrors are studied and the effects of recycling particles on the formation of axial confinement potential are discussed [29]. PWIs in the tandem mirrors TMX-U and MFTF-B with thermal barrier potentials are also discussed [30]. It is understand that presence of recycling particles has adverse effects on the formation of both axial confinement and thermal barrier potentials. End loss particles hit the end plates at end regions of a tandem mirror and causing the potential formation at end plates. End plate potential significantly controls the radial potential distribution of a mirror machine, which has significant effect on GAMMA 10 plasma parameters [31]. It is also found in GAMMA 10 experiment that end plate potential is linked with the potential at the plug regions through circulating current [32]. However, in the anchor cells of a tandem mirror, high β plasma is formed for MHD stability, where non-axisymmetric magnetic field components are present. Non-axisymmetric magnetic field components can introduce radial transport of the plasma particles [9,33]. A rigorous experimental study is carried out for the first time in GAMMA 10 to understand the plasma behavior of the anchor cell. Langmuir probes and calorimeters are used for edge plasma measurements. A part of each non-axisymmetric MFC of the anchor cells are covered by large conducting plates (APs) to control the plasma behavior. Computational study on MFC is carried out to understand the experimental results. Possible influences of the non-axisymmetric MFC and conducting plates on GAMMA 10 plasma parameters are discussed in part A of this chapter. Possible way to optimize the non-axisymmetric MFC of the anchor cell is also addressed.

Analysis of neutral particle transport is an important issue to understand the hydrogen recycling and the particle confinement in the magnetically confined plasmas [34,25]. In particular, neutral particles play an important role on the pedestal formation in H-mode plasmas and on the plasma transport in the divertor region. In the tandem mirror devices, analysis of neutral transport is also important to evaluate particle and energy confinement for obtaining high β plasmas [25,34]. In order to evaluate the neutral particle density in the hydrogen plasmas, Balmer-α line-emission from hydrogen/deuterium atoms (H_α/D_α line-emission) is measured as a standard method in various tokamaks. Neutral particle transport simulations based on the Monte-Carlo methods are widely used as a standard way to understand neutral behavior in complicated systems of fusion devices. Modeling of neutral transport using the Eirene [35] and DEGAS2 [36] Monte-Carlo codes play an important part in the studies of divertor modeling in toroidal systems, such as tokamaks. In the GAMMA 10 tandem mirror, the neutral transport code DEGAS [37] is applied for investigating the neutral particle behavior [25,38-40]. To understand the neutral particle behavior in the anchor cell, results of H_α measurements are compared with the results of the neutral transport simulation. Ambient neutral pressure effect on anchor plasma parameters is also studied by modulating gas fueling rate. Results of the experimental and computational studies on neutral particle behavior at anchor cell are also given in Part A.

Dust grains can be generated in the fusion devices by PWI through the processes such as desorption, arcing, and sputtering at the plasma facing solid materials [41-46]. In TEXTOR tokamak experiment, it is estimated that dust grains content in the machine to be over a hundred grams after long operation periods [47]. Composition of the dust grains (impurities) depends on the material of first wall and conditioning of the wall [48-53]. Most of the impurities are introduced into the plasmas in the form of atoms or molecules, but some are in the form of small solid particles such as flakes and microparticles [54]. Size of the observed solid materials spans a range from several tens of μm for microparticles to probably a few millimeters for flakes [47,55,56]. Presence of these dust grins has a crucially adverse effect on the performance of fusion plasmas [57,58]: enhanced power loss due to radiation and dilution of fuel. In mirror devices, charged dust grains formed in different mirror cells can flow along the field lines until they reach the end regions and moreover, the end loss particles hit the end plates and causing the desorption of hydrogen molecules and impurities in the end regions. Sheath parameters in front of plasma facing material can be influenced significantly due to the presence of charged dust grains [59,60]. End plate potential significantly controls the radial potential distribution of a mirror machine, which has significant effect on GAMMA 10 plasma parameters [31].Therefore, in order to correctly assess the effects of dust grains on the fusion plasma performance, some theoretical research works on the plasma with dust grains should be carried out within the range of fusion plasma parameters. Possible areas of study on fusion plasma with dust grains are charging of dust grains, collective behavior of plasma with charged dust grains, dust grains effect on the sheath properties at the plasma facing materials, etc. Due to the physical processes – such as atomic process, outgassing, etc. - low temperature and high density dusty plasma can be formed in the divertor plasma of tokamak and in the plasma near the end plates of mirror machine. Possible dust modes may explain the extremely low frequency fluctuations, new channels for the parametric coupling of other waves, generation of wake-fields, etc., in dusty plasma. Results of the theoretical study on the properties and instabilities of dust modes, such as dust-lower-hybrid and dust-acoustic modes, in a divertor plasma environment are given in part B of this chapter. Charging of dust grains is also given in this part.

2. PART A: EXPERIMENTAL STUDY ON TANDEM MIRROR EDGE PLASMAS

One of the important problems of the GAMMA 10 tandem mirror is to increase the plasma density, i.e., to improve the plasma confinement. In a tandem mirror, plasma confinement property depends on MHD stability which depends on plasma pressure profiles of different cells, radial transport processes of plasma due to non-axisymmetric magnetic or electric field configurations, resonant redial diffusion of plasma, etc. Successful axial confinement is important for further development of tandem mirror machine. The problem of increasing the plasma density by axial confinement has two folds and hence, two ways of investigation are necessary: 1) to prevent the ion losses at the ends, it is necessary to examine the basic problems of the formation and sustainment of the axial confinement potentials (plug potentials) and 2) to prevent the radial loss of the axially confined particles, it is required to examine the effect of non-axisymmetric magnetic field components on the plasma, regularity of the electric field distribution in the plasma, etc. In hot ion mode of the GAMMA 10 experiment, effective formation of axial confinement potential is found [61], but the density of the central cell plasma is not increasing enough during the axial confinement. Some radial losses are observed. The radial loss rate is measured at several percents by the one-end plugging data. Under good condition, shots with an estimated radial loss rate of less than 3% are obtained [61]. In GAMMA 10 experiment, it is also found that the plasma density increases up to certain limit. The radial loss in the axisymmetric central cell is estimated to be small. Therefore, one of the causes of insufficient increment of the central cell plasma density, especially during axial confinement, radial loss of the plasma at the highly elliptic flux tube regions of the anchor cell is being considered. In GAMMA 10, only the anchor cells have non-axisymmetric magnetic field component. We scan the plasma in vertical directions with different horizontal positions at east anchor outer transition region with a set of movable Langmuir probes. Moreover, the flux tube of the non-axisymmetric magnetic field regions is covered by the several pieces of large sized conducting plates. Experiments are carried out with floating and grounding conditions of the plates to control and to understand the local plasma behavior. Edge plasmas are also measured using calorimeters, which are installed at different positions on the APs. To understand the neutral particle behavior in the anchor cell, H_α intensities are measured and neutral transport is analyzed using the DEGAS Monte-Carlo code. To control the ambient neutral pressure in the anchor cell during axial confinement, experiments with modulating gas fueling rate are carried out. Study on MFC is done to understand the experimental results and to optimize the effect of the non-axisymmetric magnetic field. These studies have the following importances: (1) to understand the effect of non-axisymmetric magnetic field on the plasma, (2) to investigate the radial transport of plasma and the effect of the consequent plasma wall interaction, and (3) to obtain the data base to improve the design of anchor cell as well as to improve the confinement of plasma.

2-1   Experimental Set-up

2.1.1 GAMMA 10 Device

GAMMA 10 is an effectively axi-symmetrized minimum-B anchored tandem mirror [62,63], which consists of seven cells: one central cell, two anchor cells, two plug/barrier cells, and two end cells. Fig.1 shows the schematic view of the GAMMA 10 tandem mirror in which (a) is coil configuration, (b) is axial distribution of B, (c) is shape of the magnetic flux tube, and (d) is axial distribution of the plasma potential. Magnetic field configurations of the central and plug/barrier cells are simple mirror type. In the central cell main plasma is confined and in plug/barrier cells plug and barrier potentials are produced. The open and expanding magnetic lines of forces are ended to the metallic plates in the end cells. End loss particles are terminated on these plates. Each of the anchor cells is situated between the central and plug/barrier cells and consists of minimum-B MFC. High β plasma is produced in the anchor cells to provide the MHD stability of the whole plasma.

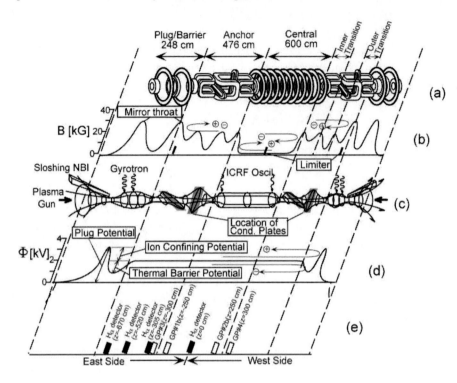

Figure 1. Schematic view of the GAMMA10 tandem mirror: (a) coil configuration, (b) axial distribution of B, (c) shape of the magnetic flux tube, (d) axial distribution of the plasma potential, and (e) locations of gas puffer and H_α detector.

The initial short pulse ($1ms$) plasmas are produced by the magneto plasma dynamics (MPD) guns which are installed in both ends of GAMMA 10. These plasmas are injected into the central cell along the magnetic field lines by the $v{\times}B$ force. A plasma is then sustained effectively with the additional gas puffing and ion-cyclotron range of frequency (ICRF) heatings and is referred as target plasma. ICRF heating (RF1:$\sim 10MHz$) is also used to heat the anchor cell ions [64]. Hydrogen molecules are introduced in GAMMA 10 as working gas through gas puffers. Locations of presently used gas puffers $\#1b$, $\#2b$, $\#3$, and $\#4$ are shown in Fig. 1 (e). Plug and barrier potentials in the plug/barrier cells are produced by the application of electron cyclotron resonance heatings (ECH) on the target plasma [65]. This mode of plasma production in GAMMA 10 is called hot ion mode. However, central cell ions are heated using RF2 ($4-10MHz$) and neutral beam injection (NBI) heating systems.

Time dependent evolution of line density and ion temperature of the plasma in the central and anchor cells are measured by microwave interferometer and diamagnetic loop, respectively. End loss ion energy spectrum is obtained by end loss particle analyzers (ELAs), which are placed at both ends of GAMMA 10. Effectiveness of plug potential and ion temperature parallel to the field line can be obtained by analyzing the energy spectrum of the end loss ions. X-ray detectors are used for the measurement of electron temperature in various regions of GAMMA 10. Total volume of the vacuum vessel of GAMMA 10 is $180m^3$. By the operation of rotary, turbomolecular, cryosorption, and liquid helium cryopanel pumping systems, gas pressure in the vacuum vessel is kept of the order of $10^{-5}Pa$.

2.1.2 Diagnostics of the Edge Plasma in Anchor Cell

Conducting plates and calorimeters

A part (about $100cm$ in z-direction) of each elliptic flux tube region of anchor cells is covered by three pairs of conducting plates (APs) $P1$, $P2$, and $P3$, as shown in Fig. 2. Locations of the APs in GAMMA 10 are shown in Fig. 1(c). Each plate of $P1$, $P2$, and $P3$ is cut into three peaces for better control and understand the local plasma behavior. The plates are fixed few cm outside the flux tube of radius 18 cm at the central cell mid-plane [66-68] and are at the high magnetic field regions of the anchor cells. Conducting plates are made of SUS 316 sheet with thickness $2mm$. Each part of the plates is electrically independent and can be biased, floated (grounded the APs with high resistance of $R_{AP}=1M\Omega$) or grounded (grounded the APs with low resistance of $R_{AP}=1\sim100\Omega$) as necessary. Several calorimeters are placed on the plates. Locations of the calorimeters on the APs at the east anchor cell are shown in Fig. 2(b). The calorimeters are the thermocouple of copper-constantan materials and are fixed on SUS sheet of thickness $2mm$ by micro-spot welding [69]. Surface area of the SUS sheet is $\sim0.5sq.cm$ and the calorimeters are placed on the APs in such a way that they are thermally insulated from the APs. Heat flux from the plasma is detected as temperature rise of the SUS plate measured with the thermocouple. Signal from the calorimeter is taken into a low speed analog recorder as an average value during a plasma shot.

Movable Langmuir Probe

The set of movable Langmuir probes consists of two probes in such a way that one probe can be rotated circularly with respect to other with radius $7cm$ [9]. The probe along the axis of rotation is y-probe and the other is x-probe. The axis of rotation of the probe system is placed at the transition region of $z=-667cm$ as shown in Fig. 2(c). This area is about $27cm$ far away from the edge of the conducting plates $P1$. The positions of the probe tips in the plasma are determined by mapping field lines. Field lines of different y_{cc} are also shown in Fig. 2(c). y_{cc} is the distance of the field lines on y axis at the central cell mid-plane. Langmuir probes are made of tungsten wire. Ceramic pipe (which is used to cover the tungsten wire) is made as thin as possible and also the length of the pipe is determined to be long enough. Therefore, the edge plasma is not disturbed due to the presence of ceramic pipe and the probe holder during the measurement at our interested locations inside the plasma. Probes are biased with a triangular shaped sweep voltage of $500V$ and a frequency of $100Hz$ in order to get the I-V characteristics of the plasma at every $10ms$ interval. In analysis of I-V characteristics, probe current only during the rising part of the probe voltage is considered.

H_α detectors

H_α detectors are installed in the several positions from the mid-plane of the central cell to the east anchor cell in order to investigate the neutral particle behavior in this region. Locations of the H_α detectors are shown in Fig. 1(e). Each detector consists of an H_α

interference filter (band width at full width at half maximum $\Delta\lambda_{1/2}$ $=2.7nm$), an optical fiber, and a photomultiplier. Three sets of detector are installed at the anchor mid-plane and at the inner and outer transition regions. The detected H_α light is transferred to magnetically shielded photo-multiplier via optical fiber and the converted electric signal is then moved to the CAMAC system and finally analyzed with a workstation. Profile of neutral density is determined with a help of neutral transport simulation.

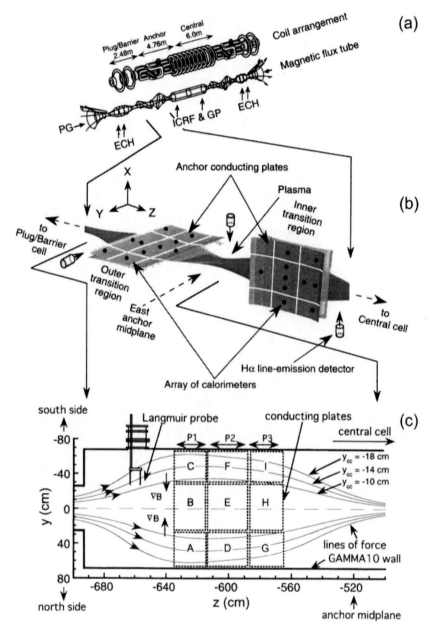

Figure 2. Schematic diagram of (a) coil arrangement, shape of the magnetic flux tube, and location of heating systems of GAMMA 10 and (b) APs at east anchor cell and locations of calorimeters on APs and H_α detectors. The movable Langmuir probe at the outer transition regions of east anchor cell, magnetic lines of forces in yz-plane, and position of APs are shown in (c).

2-2   Anchor Cell of GAMMA 10
2.2.1   Geometry of the Plasma in Anchor Cell

Minimum-B magnetic field has strong MHD stabilization property [70] and this field can be produced using many types of coils, such as hexapole, quadrupole, baseball, and Yin-Yang coils [71]. Minimum-B is the basic concept to stabilize the plasma in all mirror confinement systems. In a tandem mirror, minimum-B anchor cells are combined with axisymmetric mirror cells [72,65,73]. Minimum-B in the anchor cells of the GAMMA 10 tandem mirror effectively reduces the MHD instability [74]. In GAMMA 10, minimum-B configuration is connected with the neighboring mirror cells through the transition regions, where the MFC is non-axisymmetric and the circular flux tube of the axisymmetric regions becomes highly elliptic. The two transition regions of an anchor cell have $\pi/2$ symmetry and hence, the directions of the curvature and ∇B drifts of the particles in inner transition are opposite to the directions in outer transition region as shown in Fig. 3. Direction of the curvature and ∇B drifts is given by $\hat{y} = \pm(\bar{R}_c \times \bar{B}/|\ \bar{R}_c \times \bar{B}\ |)$, where R_c is the radius of curvature, $\boldsymbol{B}$ is the magnetic field, $'+'$ sign is for ions, and $'-'$ sign is for electrons. Hence, in the absence of azimuthal rotation of plasma (due to diamagnetic drift, $E \times B$ drift, etc), the non-axisymmetric magnetic field effect is thus thought to be canceled out for the particles which pass through these two regions. To allow the bounce of central cell particles through the anchor cells, magnetic or electrostatic reflection of the particles is prevented to occur in these cells. Moreover, to avoid the resonant radial diffusion, the magnetic flux cross-section is made circular in any region where central cell particles are reflected magnetically or electrostatically. The necessity of axi-symmetrization is that the non-axisymmetric components may cause resonant radial diffusion which eventually limit the confinement time of tandem mirror central cell ions [75].

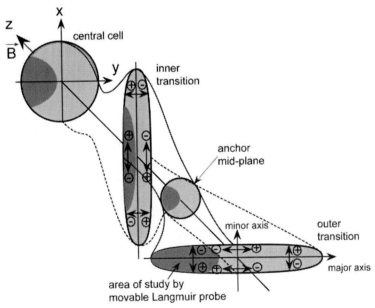

Figure 3. Directions of curvature and ∇B drifts of the particles are shown in the cross-sections of the flux tube at the inner and outer transition regions.

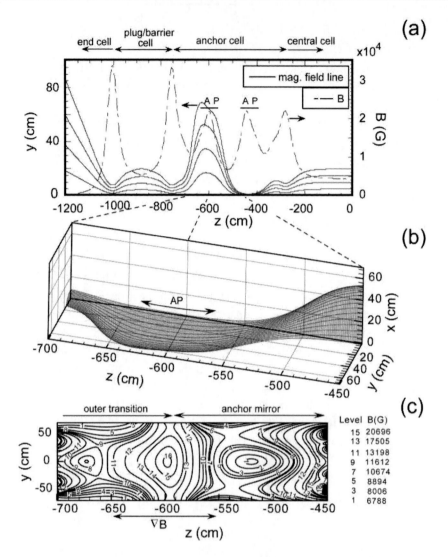

Figure 4. (a) Half of the axial distribution of **B** of the GAMMA 10 and field lines in yz-plane. (b) Shape of the magnetic flux tube of $r_{cc}=18\,cm$ in the anchor cell region, where a quarter of the flux tube in the azimuthal direction is shown. (c) Mod-B contour in yz-plane at the anchor cell.

From the calculation of magnetic field lines, as shown in Fig. 4(a) by solid line, it is seen that the radius of curvature of a field line in minor axis side of the elliptic flux tube is larger than that of in major axis side. This difference of radius of curvatures of a field line increases with the increase of radial positions of the field lines. Length of each transition region is $165\,cm$ and has mirror magnetic field configuration of mirror ratio two as shown in Fig. 4(a) by dotted line. Magnetic flux tube around the transition regions of each anchor cell is highly elliptic as shown in Fig. 4(b). Shape of the flux tube at one mirror reflection side of the transition region is almost circular, but that of the other mirror reflection side is highly elliptic, i.e., reflection points of the mirror transition regions have $\pi/2$ asymmetry as shown in Fig. 4(b). Therefore, the magnetically confined particles in the transition regions will suffer $\pi/2$ non-axisymmetric resonant diffusion in each reflection. However, the two mirror reflection sides of the central minimum-B region have $\pi/2$ symmetry. In Fig. 4(c), contour of magnetic field of the anchor cell in yz-plane is shown. Minimum-B regions at the anchor mid-plane and at outer transition region can be seen from this figure. Between these two minimum-B regions there is about $100cm$ (from $550cm$ to $650cm$) long positive ∇B region where the anchor conducting plates are installed.

2.2.2 Estimation of Particle Shift Due to ∇B and Curvature Drifts

Magnetic field configuration of each of the four transition regions of the GAMMA 10 has 100 cm long non-axisymmetric positive ∇B. During the time of particle flow through the each ∇B region, particles will drift gradually in perpendicular direction with their drift velocities and the amount of perpendicular shift $\bar{d}_\perp$ can be written as:

$$(\bar{d}_\perp)_{i,e} = \left(\left\langle \bar{v}_{R_c} + \nabla B \right\rangle\right)_{i,e} \times (t_{||})_{i,e}, \qquad (2.1)$$

where the subscripts i and e indicate the ion and electron quantities, $\left\langle \bar{v}_{R_c} + \nabla B \right\rangle$ is the average drift velocity due to ∇B and curvature drifts and is given by according to the single particle theory [76]:

$$\left(\left\langle \bar{v}_{R_c} + \nabla B \right\rangle\right)_{i,e} = \pm \frac{(\langle r \rangle)_{i,e}}{R_c}(C)_{i,e}\hat{y}. \qquad (2.2)$$

Quantity $t_{||}$ is the time required for particles to pass the distance s. s is the path length of the ∇B region in each transition region side of an anchor cell. $t_{||}$ is given by:

$$t_{||i} = \frac{s}{v_{||i}} \text{ and } t_{||e} = \frac{s}{C_e}. \qquad (2.3)$$

The quantity $\langle r \rangle$ is the average Larmor radius, C_j is the thermal velocity of the particle species j, $\bar{R}_c$ is the radius of curvature, $\hat{y}$ is the direction of the drift which is given by $\hat{y} = \dfrac{\bar{R}_c \times \bar{B}}{|\bar{R}_c \times \bar{B}|}$, and $v_{||}$ is the parallel velocity of the particles. Expressions for electron and ion Larmor radii are $r_i = 1.44\times10^{-4}\sqrt{T_i}/B$ and $r_e = 3.37\times10^{-6}\sqrt{T_e}/B$, respectively, where $T_{i,e}$ are the specific particle temperature. Thermal velocity of ion and electron can be calculated from the expression $C_i = 1.38\times10^4\sqrt{T_i}$ and $C_e = 5.93\times10^5\sqrt{T_e}$, respectively.

Passing time ($t_{||}$) and perpendicular shift ($d_\perp$) of the species can be calculated using known values of: $|\bar{B}|=1.3T$, $|\bar{R}_c|=0.5m$, $T_e=100eV$, $T_i=T_{||i}=200eV$, $T_{\perp i}=300eV$, and $s=1m$. The passing times $t_{||i}=5.13\times10^{-6}sec$ and $t_{||e}=1.69\times10^{-7}sec$ are obtained. In these times, ion will shift $d_{\perp i}=0.4cm$ and electron will shift $d_{\perp e}=0.005cm$ in perpendicular direction with their drift velocities. Schematic view of the particle shift in a transition region is shown in Fig. 5(a). Therefore, ion will be affected seriously by the non-axisymmetric positive ∇B in the anchor cell due to the vast difference of parallel velocity and Larmor radius of passing ions and electrons. Moreover, shifting of ion position depends on the anisotropy of the ion velocity, i.e., on the ratio $(v_\perp/v_{||})^2$. Ions with the larger velocity anisotropy will be affected more by the non-axisymmetric ∇B. According to the Eq. (2.2), the direction of $\overline{v}_{R_c+\nabla B}$ for ion in the side of study ($-y$ side) by movable Langmuir probe is downward ($-x$) while that for electron is upward ($+x$) as shown in Fig. 5(b).

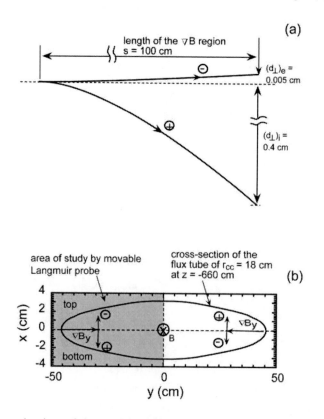

Figure 5. (a) Schematic view of the particle shift due to the curvature and ∇B drifts and (b) Cross-section of the flux tube at the outer transition region and explanation of the drift direction of ion and electron in xy-plane.

The ∇B region is around the high magnetic field area of maximum field strength $2T$. Mainly, passing particles will be affected by this non-axisymmetric ∇B configuration. Positive ambipolar potential is the natural consequence of mirror machine due to the disparity in ion and electron velocities and this potential modifies the loss cone configuration [77]. The effect of the ambipolar potential on the ion is that low energy ions are expelled, only ions that exceed a minimum kinetic energy can be mirror confined. Moreover, this potential has a typical value of four to five times of $k_B T_e$, so that most of the electrons are confined by the ambipolar potential, i.e., only electrons in the high energy tail of the distribution function escape over the barrier. Fig. 6 shows the modification of loss cone by ambipolar potential in velocity space (for the velocity distribution at the mid-plane of a mirror cell) in which (a) ion loss cone and (b) electron loss cone. Therefore, in the presence of ambipolar potential, ions (slower particles) will be shifted more compared to electrons (faster particles) during their flow through the non-axisymmetric magnetic field regions of the anchor cells as given in Eq. (2.1).

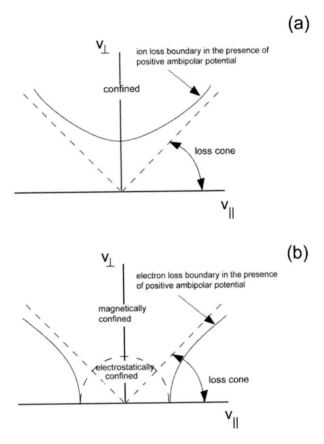

Figure 6. Modification of loss cone by ambipolar potential in velocity space in which (a) ion loss cone and (b) electron loss cone.

2-3   Results and Discussions of Edge Plasma Measurement in Anchor Cell

2.3.1 Measurement by Movable Langmuir Probe

Plasma at the outer transition region of east anchor cell (at $-z$ and $-y$ sides) is scanned radially and horizontally by a set of movable Langmuir probes as shown in Fig. 7(a). Measurements are carried out during the hot ion mode of the GAMMA 10 operation [12]. We studied the plasma structure, in various conditions of the conducting plates (grounding and floating), different positions of movable limiters, and various powers of RF and plug ECH. In any case, significant asymmetry of ion and electron saturation currents in the minor axis direction of the elliptic flux tube is observed. Direction of the shifting of plasma has no relation with plug ECH and the movable limiter positions. Movable limiters are situated at $z = \pm 706$ cm. Probe currents (ion and electron saturation currents) in the minor axis direction at $y_{cc}=-14cm$ in the case of grounded anchor conducting plate ($R_{AP}=100\,\Omega$) and with the movable limiter at $r_{cc}=12.3$ cm are shown in Fig. 7(b). y_{cc} indicates the position of magnetic field lines on y axis at the central cell mid-plane. r_{cc} presents the radial distance of the field lines at the central cell mid-plane. NBI and barrier ECH heatings are not used during this experiment. The RF1, RF2 and plug ECH heating powers are $68kW$, $100kW$, and $120kW$, respectively. As shown in the figure, the ratio of the ion saturation current to the electron saturation current is much larger than the standard Langmuir probe theory. This is considered to come from the high ion temperature compared to the electron temperature by nearly one order of magnitude and from the high magnetic field as illustrated in Fig. 4. Electron temperature in the anchor cell has been obtained to be ~$50eV$ from soft X-ray measurements and that of ions has been measured to be ~$500eV$ using a time-of-flight type neutral particle analyzer. Under these circumstances, however, larger ion and electron saturation currents are observed in the bottom side compared to the top side. Increment of electron saturation current is observed both in the top and the bottom sides during axial confinement (plug ECH). Ratio of the ion saturation currents between two corresponding points in the bottom (e.g. $x=-1cm$) and the top (e.g. $x=1cm$) sides, $\left[\left(I_{sat}^{+}\right)_{x=-1,-2}\Big/\left(I_{sat}^{+}\right)_{x=+1,+2}\right]$, is plotted against y_{cc} as shown in Figure 7(c). x axis is in the minor axis direction. This figure shows that the top bottom asymmetry of the saturation currents increases in outward direction.

Radius of curvature of the field lines and the magnetic field strength in minor axis side of the elliptic flux tube are larger than those in major axis side. Therefore, drift velocity of the particles in minor radius side will not be same as the drift velocity of the particles in major radius side [cf. Eq. 2.1]. In addition, diamagnetic drift is a natural consequence of the mirror confined plasma due to the presence of plasma pressure gradient and is observed in GAMMA 10 [78]. In the presence of azimuthal rotation of plasma (due to diamagnetic drift, $E \times B$ drift, etc), the effect of non-axisymmetric magnetic field component on the bounced particles will become crucial. Moreover, passing particles from the anchor mirror cell will be affected by the non-axisymmetric positive ∇B either at the inner transition or at the outer transition regions. Under these circumstances, there are two possible mechanisms to explain the probe current asymmetry in minor axis direction as follows: 1) Due to the vast difference in Larmor radius and parallel velocity of ions and electrons, ions will be strongly affected by the non-axisymmetric magnetic field compared to electrons under the present experimental conditions (as discussed in 2.2.2). While electron will act as a charge neutrality fluid due to the smaller moment of inertia. Hence, the plasma as a whole will shift to ion drift side. 2) By the collision of the drifted out ions with the wall of vacuum vessel or with the grounded conducting plates, neutral particles will come off from the wall/plate. The enhanced outgassing from the wall/plate in the ion drift side will cause the plasma density to increase because of the ionization of recycling neutrals.

Reason of the gradual increase of the top bottom asymmetry of probe currents (Fig. 7(c)) in outward direction is explained as follows. Normal curvature increases in major radius direction of the elliptic flux tube - because the curvature is proportional to the radius - and the magnetic field strength decreases in this direction. Hence, the particles in the outer region will be affected more by the ∇B and curvature drifts compared to the particles in inner region of the elliptic flux tube. However, the mechanisms as explained in (1) and (2) may have separate role or combined role to the probe current asymmetry in the outer transition region. In any way, the non-axisymmetric ∇B and curvature drifts are the main sources of plasma shift, i.e., the plasma flow in the non-axisymmetric magnetic field region is responsible for the probe current asymmetry.

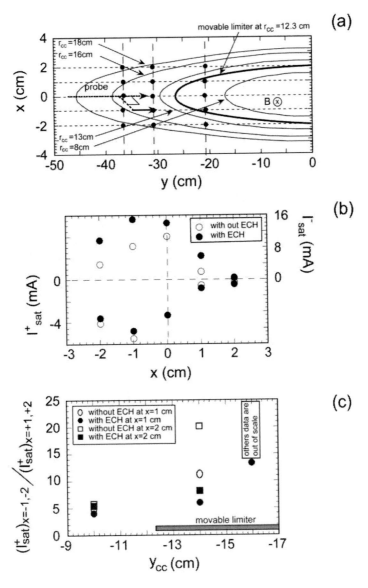

Figure 7. (a) Description of the area of study by movable Langmuir probe. Cross-section of the flux tube of different r_{cc} in xy-plane at $z=-660\,cm$ and the position of the x-probe tip in the plasma during scanning is shown by solid circles. (b) Ion and electron saturation currents measured by x-probe in the vertical direction at $y_{cc}=-14\,cm$. (c) Top bottom asymmetry of ion saturation current at different positions in horizontal plane.

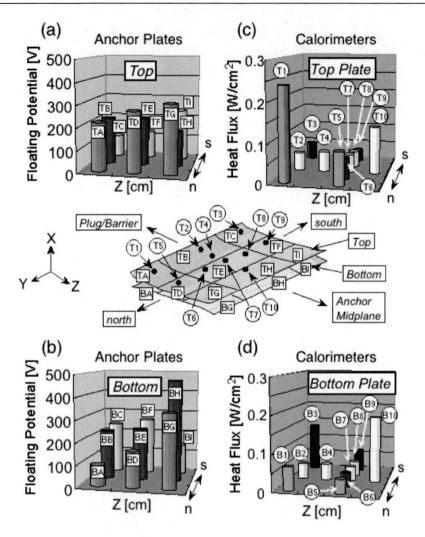

Figure 8. Spatial profiles of the floating potentials of APs and heat flux measured by calorimeters at the outer transition region of the east anchor cell. Characters enclosed by squares and circles represent the location of anchor plates and calorimeters, respectively.

2.3.2 Measurement by AP

Floating potentials of the conducting plates (APs) are measured in the hot ion mode of the GAMMA 10 operation. In the floating condition of the APs, potential of the plates arises from the net electric charge generated at the plates by the difference between the electron and ion fluxes in the direction of the plate surface from the plasma. Floating potentials of the APs at the outer transition side of east anchor cell during without plug ECH, i.e., when the plasma is sustained with RF waves are given in Figs. 8(a) and 8(b). All the plate potentials are positive. Maximum floating potentials of the plates at the top and bottom sides are about *300V* and *450V*, respectively. Floating potentials of the plates T_A, T_B, and T_C of *P1* and T_D, T_E, and T_F of *P2* at the top side have a tendency to increase form south side (s) to north side (n) as shown in Fig. 8(a), but that of the plates of *P1* and *P2* at the bottom side are decreasing in this direction as shown in Fig. 8(b). In other words, the floating potentials of the plates in ion drift side are larger. The measurements by *P1* and *P2* agree well with the ion loss mechanism by ∇B and curvature drifts. Positions of the plates of *P3* are inside the mirror field of magnetic-well, where the magnetic field configuration is comparatively axisymmetric. Therefore, the floating potentials of the top and bottom plates of *P3* do not show the clear effect of ∇B on plasma.

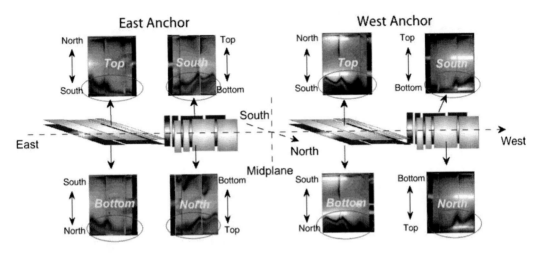

Figure 9. Photograph of the anchor conducting plates (plasma facing surface) after using three years. The circles show the area of deposited impurity material.

APs are at high magnetic field region, where the flow of passing particles is present. Therefore, the positive floating potentials of the anchor plates can be thought to be due to the interaction of the drifted out passing ions with the plates. Impurity deposited areas are observed on the APs as shown in Fig.9. In Fig. 9, all APs in the inner and outer transition regions of east and west anchor cells are shown after three years operation (*1999–2001*). During this period, flux tube of each transition region was covered by four pairs of AP. Impurity deposited areas are in the opposite side of the ion drift due to ∇B and curvature drifts as shown in Fig. 9. This observation indicates that sputtered impurities from the plates of ion drift sides are redeposited on the plates at the opposite sides of ion drift.

2.3.3 Measurement by Calorimeter

Heat flux from the plasma is measured by the calorimeters and the measured heat fluxes at the outer transition region are shown in Figs. 8(c) and 8(d). These measurements are carried out in the similar plasma shots that was for the measurement of APs potentials. Distributions of heat fluxes from south to north sides are exactly consistent with the distributions of floating potentials of the APs as discussed above, i.e., the heat flux is larger in ion drift side. Therefore, this agreement between heat flux and positive potential of AP indicates that the heat flux is dominated by ion flux. Measurement by calorimeters is also carried out during the plasma shot with plug ECH. It is observed that the direction of asymmetry in the heat flux is independent of plug ECH, i.e., axial confinement. These observations indicate that the asymmetry in the measurement by calorimeters is due to the non-axisymmetry magnetic field effect on the plasma particles.

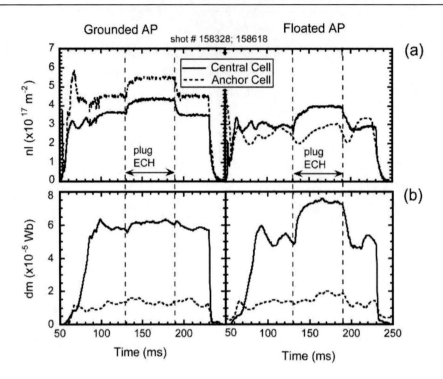

Figure 10. (a) Line density and (b) diamagnetism of the central and anchor cells in the case of floated and grounded anchor conducting plates (APs).

2.3.4 Plasma Parameters of GAMMA 10

Time dependent line densities (nl) of the plasma in the central (cc) and east anchor (ea) cells are shown in Fig. 10(a), for both grounding and floating conditions of AP. In Fig.10(b), diamagnetisms (dm) of the plasma in the central and east anchor cells for both the cases of AP are given. Solid and broken lines represent the data in central and anchor cells, respectively.

Data for the case of grounded AP indicate that in the anchor cell comparatively low temperature, but high density plasma is formed compared to the central cell plasma. In this case of AP, nl_{cc} increases without significant increasing of dm_{cc} during plug ECH. On the other hand, nl_{ea} increases, but dm_{ea} decreases during plug ECH. These results can be explained as follows. The desorped gas due to collision of the drifted out ions with the grounded APs will be ionized and will be trapped in the anchor cells (as discussed in 2.3.1.). Temperature of the plasma will decrease by the charge exchange process. The fast neutrals due to charge exchange process will then collide with the plates and desorption will occur again. This process will continue like a chain reaction and higher density and lower temperature plasma may form in this way in the anchor cell. ∇B and curvature drift velocity directions are independent of the particle velocity direction. Hence, the passing ions those bouncing between the two plug potentials will be drifted out the flux tube easily. During axial confinement wall plasma interaction will increase due to the increase of radial loss. Therefore, during plug ECH, further increasing of plasma density and decreasing of plasma temperature may occur. High density, i.e., high potential plasma (according to Boltzmann relation: $\phi = k_B T_e / e \ln (n/n_o)$, where ϕ, k_B, T_e, e, and n are the plasma potential, Boltzmann constant, electron temperature, charge of electron, and plasma density, respectively) in the minimum-B region compared to that of central cell plasma can be produced. Electrostatic reflection of passing ion from central cell by the potential at anchor cell can introduce resonant diffusion of the ion. Resonant diffusion of passing ion can limit the increment of the plasma density in central cell [75].

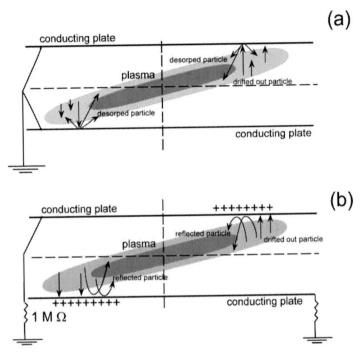

Figure 11. Schematic view of plasma shape in the transition regions and possible PWI in the case of (a) grounded and (b) floated AP. Electrostatic reflection of the drifted out particles from the floated APs due to the self biasing effect is shown.

Data for the case of floated AP indicate that in the anchor cell comparatively low temperature and low density plasma is formed compared to the central cell plasma. In this case, both the plasma density and diamagnetism in central and anchor cells increase during plug ECH as shown in Fig. 10. Simultaneous and significant increment of both the plasma density and diamagnetism at different cells during plug ECH is observed for the first time in GAMMA 10 experiment. It is thought that due to the self biasing effect of floated APs, floated APs can reduce PWI: at first the APs will become positively charged due to the collision of fast drifted out ions. Subsequently, drifted out ions with lower velocity will be reflected electrostatically by the positively charged conducting plates. Hence, the PWI will reduce. Small value of nl_{ea} during floated AP compared to the case of grounded AP is due to the decreasing of desorbed neutrals. Though, diamagnetism depends on the plasma density, large value of dm_{cc} with small value of nl_{cc} is observed in the case floated AP compared to the case of grounded AP, especially during plug ECH. Therefore, in the case of floated AP comparatively hot plasma is formed. According to the explanation of the experimental results, schematic diagrams of the plasma shape at the transition region of the anchor cell and plasma wall interaction in the case of grounded and floated APs are shown in Fig. 11.

In the absence of collision, passing particles can not be trapped in the mirror field of the transition regions, because of the conservation of magnetic momentum. On the other hand, the plasma that formed due to the ionization of desorbed neutrals in the transition region can be trapped in the mirror field of the region. These magnetically confined particles in the transition regions will suffer resonant diffusion associated with uncanceled azimuthal drift $\sim\pi/2$ on each reflection (as discussed in 2.2.1). Drifting out of ions from the transition regions may be good for the MHD stability [79], but recycling of neutrals due to PWI may prevent the formation of high β plasma in the anchor cell. Therefore, floated AP helps to from high β plasma in the anchor cells by reducing PWI in the way of self biasing effect of the AP. According to the experimental results, a model of the particle behavior in the anchor cell is made as shown in Fig. 12. Magnetically and axially confined particles and their drift in radial direction due to the effect of non-axisymmetric magnetic field in the transition regions are shown.

Pressure of additional amount of desorbed neutrals during plug ECH can be controlled by decreasing gas fueling rate. Comparative study of the plasma parameters including H_α measurements at the anchor and central cells between the cases without and with gas fueling rate modulation is carried out. Result of this study is given in section 2-5 after the detailed discussion of neutrals behavior in the GAMMA 10, especially in the anchor cell in section 2-4.

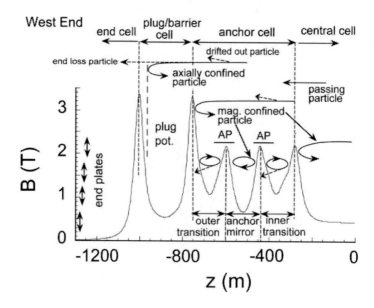

Figure 12. Model of plasma particle behavior in GAMMA 10.

2-4　Neutral Particle Behavior in the GAMMA 10

2.4.1　H_α Measurement in Central Cell

In the GAMMA 10 tandem mirror, neutral hydrogen density in the central cell are estimated by measuring spatial-profiles of H_α line-emission with axially and radially aligned

H_α detectors [25,38,39,80,81]. These detectors are absolutely calibrated by using a standard lamp, which enables us to evaluate the neutral density from the absolute value of the H_α emissivity based on the collisional-radiative model [82]. Computer simulations on the neutral particle transport using the DEGAS Monte-Carlo code [37] are carried out to understand the experimental results. The code has been modified to take into account of dissociative-excitation reactions of molecular hydrogen in order to adapt the simulation to low density range [25,83].

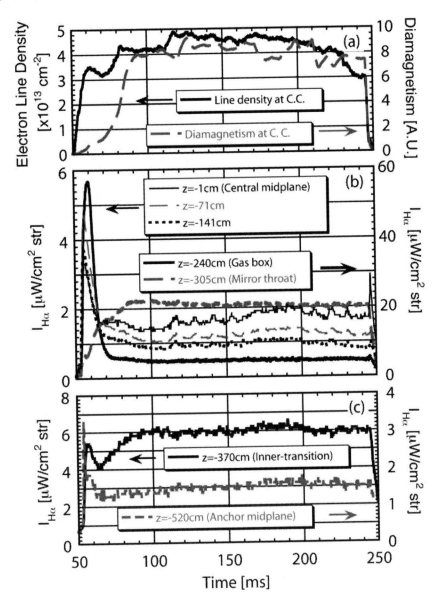

Figure 13. (a) A typical time history of GAMMA 10 plasma parameters during ICRF heating. (b) Intensity of H_α emission measured at the central cell, near the gas box, and mirror throat. (c) Intensity of H_α emission measured at the anchor cell.

In Fig. 13, a typical time history of plasma parameters and of H_α line emission is shown in the ICRF-heated GAMMA 10 plasma. In this experiment, two types of gas puffers are used. One (GP#1b) is utilized for start-up of the plasma by injecting dense gas into the gas box ($z=-240cm$) with a short pulse. The other (GP#3a) is used for sustaining the plasma along with the ICRF pulse by continuously introducing gas in the mirror throat region ($z=-300cm$). Fig. 13 (b) shows the intensity of H_α emission measured with the axial detector array of the central cell and those measured at the gas box and the mirror throat. At the initial phase of plasma start-up, the signal at the gas box has a strong peak. This indicates that the influence of the gas puffer GP#1b is dominant in the initial phase of plasma start-up. In a steady state phase, on the other hand, the H_α intensity from the mirror throat is stronger than the other region, which implies that hydrogen atoms induced from GP#3a are localized in the vicinity of the mirror throat. The brightness of H_α emission detected at the mirror throat is more than ten times larger than that near the mid-plane of the central cell.

2.4.2 H_α Measurement in Anchor Cell

The intensity of H_α emission measured at the anchor cell for the first time [84] is shown in Fig. 13(c). The observed peak in the initial phase also indicates the particle feed from the gas box and the plasma production in the anchor cell by the RF1 heating. After $80ms$, the time behavior of H_α emission observed at the inner transition region is different from that measured at the anchor mid-plane. The time evolution in the inner transition is similar to that of the mirror throat. These characteristics may be explained from the mechanism that the plasma escaping from the mirror throat can smoothly flow into the transition region, since there is no particle trapping function in this region. On the other hand, the observation that the signal in anchor mid-plane resembles to that of the central cell is ascribed to the mechanism that the passing particles in the anchor mid-plane are heated and trapped in the resonance area of RF1 which has the same as the central mid-plane.

2.4.3 Neutral Transport Simulation using DEGAS Monte-Carlo Particle Code

In order to execute the three dimensional neutral transport simulation in anchor cell, a modeling of the simulation space for the anchor cell is performed. The mesh model of the wall-surface of the anchor minimum-B region and the surface plot of the plasma grid are shown in Fig.14. As shown in the figure, the model introduces symmetry in vertical direction and the simulation space is divided into 11 segments radially and 8 segments azimuthally. In the axial direction, 41 segments are utilized for the simulation. In this simulation, gas desorption near the movable limiter located in the outer transition region of the anchor cell is considered together with the particle source from the gas puffer near the mirror throat region.

Fig. 15 shows the $3-D$ simulation results of neutral hydrogen density profiles. In the anchor mid-plane, as shown in Fig. 15(a), it is recognized that the molecular hydrogen density radially decreases toward the core region of the plasma and further reduces along with the magnetic field line from the mirror throat ($z=-300cm$) to the anchor mid-plane ($z=-520cm$). Penetration of neutrals is taking place in the edge region where the plasma thickness becomes thin. The result implies that injected hydrogen molecules from the mirror throat gas puffer are transported to the anchor transition region with significant attenuation and only a small amount of them reaches the anchor mid-plane.

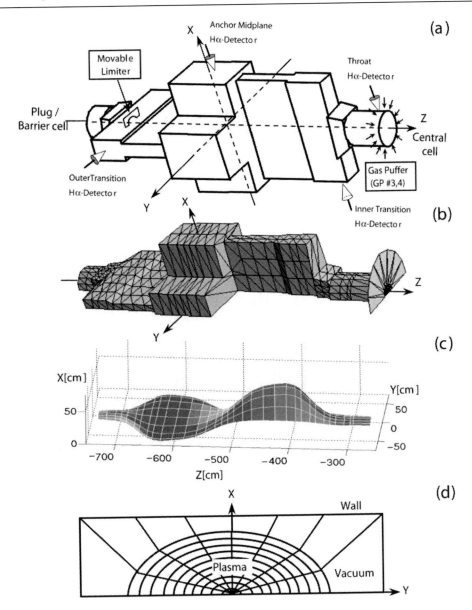

Figure 14. (a) Mesh model of the anchor cell used for DEGAS. Configuration of plasma facing walls at the anchor cell. (b) Mesh model of the wall-surfaces at the anchor cell. (c) Surface plot of the plasma grid at the anchor cell. (d) Cross-section of the mesh divided into 11 segments radially and 8 segments azimuthally.

In contrast to the molecular density, atomic hydrogen density shown in Fig. 15(b) has a tendency to concentrate in the plasma core region. The atomic density also shows the rather uniform density over the plasma cross section in inner and outer transition regions where the plasma cross section becomes much elongated elliptically. Although these mechanisms are not clarified in detail, longer mean free path of Franck-Condon neutrals compared with the plasma thickness may enhance the production of atomic hydrogen in inner region of the plasma column and form such density profiles in the anchor cell.

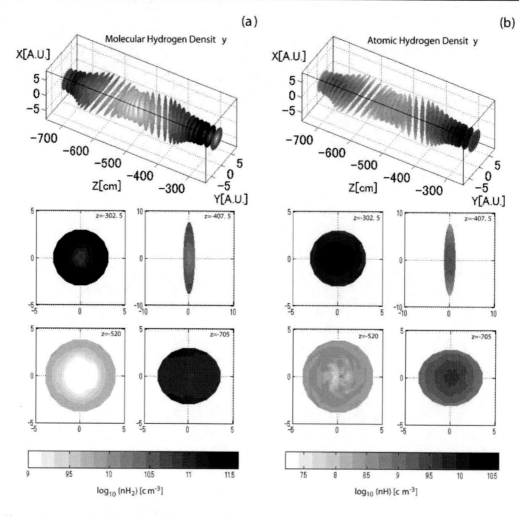

Figure 15. *3-D* simulation results of neutral hydrogen density profiles: (a) for molecular hydrogen and (b) for atomic hydrogen.

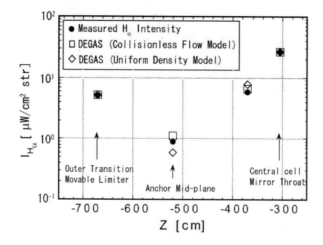

Figure 16. Comparison of the simulation results with the measured results of H_α intensity.

Fig. 16 shows the comparison between the simulation results and measured one. Calculated results of H_α intensity obtained with two typical density models are normalized so as to fit to the measured results at each source location. The simulation assuming both particle sources from the gas puffer and recycling near the movable limiter of the outer transition region successfully reproduces the axial profile of measured H_α intensity. While, a small discrepancy is observed at the anchor mid-plane in the case of the uniform density model. From the above results, it is understand that the hydrogen recycling and particle penetration in the anchor transition regions play an important roll on the behavior of neutral particles in the non-axisymmetric anchor region.

2-5   Control of Ambient Neutrals in Anchor Cell

To form high β plasma in the anchor cell, it is important to keep high recycling neutrals away from the plasma, because of the cooling associated with the charge exchange, ionization, etc. Due to the presence of additional amount of drifted out particles during axial confinement (plug ECH), recycling of neutrals will increase as explained in 2.3.4. The effect of the additional amount of recycling neutrals during axial confinement can be suppressed by the decreasing of gas fueling rate. Experiments with gas fueling rate modulation during plug ECH are carried out to understand the neutral effect on anchor cell plasma.

2.5.1   Results and Discussions of Neutral Control experiment

Gas fueling rate of the gas puffers #3 ($z=-305cm$) and #4 ($z=305cm$) near the central cell mirror throats (as shown in Fig. 1(e)) is modulated at the time of plug ECH. In the present experiment, gas pressure in the reservoir of the gas puffers #1b and #2b is $220Torr$ and that of the gas puffers #3 and #4 is $60Torr$. Gas puffers #1b and #2b are operated for the duration of $40-60ms$ to build up the initial plasma and that for gas puffers #3 and #4 is $55-240ms$ to sustain the plasma. Experiments are carried out in the hot ion mode of operation, when RF2, RF1, and Plug ECH powers are $56kW$, $75kW$, and $140kW$, respectively. The APs are in floating condition and in the phase of surface conditioning. The H_α measurements at $z=-305cm$ (near the gas puffer #3), at $z=-520cm$ (mid-plane of anchor cell), and at $z=-670cm$ (outer transition region) during without and with modulation are shown in Fig. 17. In this case, gas fueling rate of the gas puffers #3 and #4 is decreased by 20%. To coincide the period of gas fueling rate modulation with the period of plug ECH, we set the applying time of voltage at the piezoelectric valve for modulation $5ms$ before the both on and off times of the plug ECH. This time delay is due to the conductance of gas flow and slow response of gas puffing system.

Desorption of gas from the floated APs during plug ECH is observed in the phase of insufficient surface conditioning of APs. It is seen from the Fig. 17 that during gas fueling rate modulation, H_α intensity decreases significantly (36.7%) near the gas puffer and the increment of H_α intensity at the outer transition region due to recycling neutrals is almost suppressed. It is also seen that the H_α intensity at anchor mid-plane decreases during modulation. Fig. 18 shows the line density and diamagnetism of the central and east anchor cells plasma during (a) without and (b) with the modulation in the case of floated APs. Both the line density and diamagnetism of the central cell plasma are higher compared to that of anchor cell in any case with and without plug ECH and gas fueling rate modulation. Due to the ionization of recycling neutrals, the diamagnetism of the anchor cell plasma is decreasing gradually during plug ECH in the case of without modulation, as shown in Fig. 18(a). In this case, the line density of the anchor cell is increasing during plug ECH due to the combined effects of axial confinement of passing particles and mirror confinement of ionized neutrals. On the other hand, in the case of modulation, the diamagnetism of the anchor cell plasma increases 15% during plug ECH, as shown in Fig. 18(b). In this case, the increment of the anchor cell plasma density during plug ECH is 3.2% small compared to the case of without modulation. This decreasing of plasma density is thought to be due to the decreasing of neutrals. It is observed that H_α intensity near central cell mid-plane depends on flow rate of puffing neutrals and has no significant effect of desorbed neutrals from APs on this intensity due to the far distance of APs from this location of measurement.

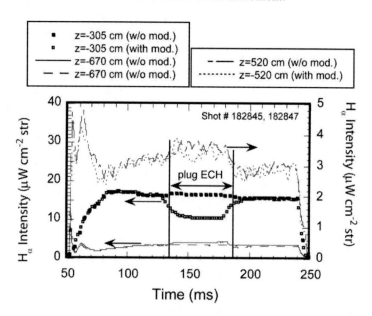

Figure 17. H_α intensities at $z=-305cm$ (near the gas puffer #3), at $z=-520cm$ (anchor mid-plane) and at $z=-670cm$ (outer transition region) during without and with gas fueling rate modulation at the time of plug ECH.

From the three dimensional neutral transport simulation using the DEGAS Monte-Carlo code, it is found that neutrals from the gas puffers #3 and #4 transport to the inner transition region with significant attenuation and only a small amount of them reaches the minimum-B region [cf. section 2-4]. Over all decreasing of H_α intensity in the anchor cell is observed during modulation. It is also observed that instead of decrease of dm_{ea}, dm_{ea} increases significantly with nl_{ea} during gas fueling modulation. This investigation indicates the presence of enhanced ambient neutrals in the anchor cells during plug ECH. Comparatively small increment of nl_{ea}, nl_{cc}, and dm_{cc} by potential confinement is found during modulation. Smaller increment of nl_{ea} and nl_{cc} is due to the decreasing of neutrals due to the modulation. Diamagnetism depends on the plasma density and hence, smaller increment of dm_{cc} is consistent with the smaller increment of nl_{cc}. Movable limiters at OT are usually placed radially more inside the plasma to control the plasma disruption related with plug ECH. Radially lost particles from the flowing plasma are collected by these limiters before going to the GAMMA 10 wall/APs. Therefore, possible cause of the decreasing of H_α at OT during modulation is the decreasing of plasma flow, which is consistent with the smaller value of plasma densities. The possibility of resonant radial diffusion of the bounced particles due to the electrostatic reflection at the anchor cell plasma is negligible, because of the lower density and diamagnetism of the anchor cell plasma. Hence, the additional neutral pressure can be considered as due to the presence of recycling neutrals, which are produced through the interaction of the drifted out particles (due to the non-axisymmetric magnetic field effect) with the GAMMA 10 wall/APs.

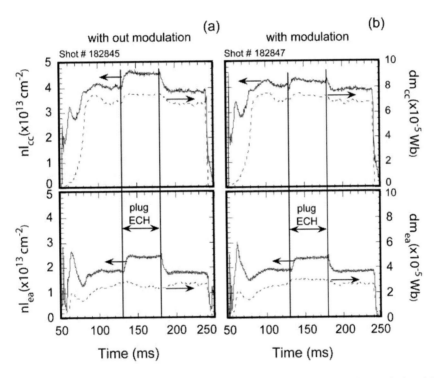

Figure 18. Line density and diamagnetism of the central and east anchor cells plasma during (a) without (b) with gas fueling rate modulation at the time of plug ECH.

2-6   Optimization of Non-axisymmetric MFC of the Anchor Cell

To ensure the MHD stability of the tandem mirror plasma using the minimum-B configuration, high β plasma is formed in the minimum-B region so that the pressure weighted average good curvature becomes larger than the pressure weighted average bad curvature of the machine. Minimum-B configuration can be obtained using baseball coils, where the components of current perpendicular to the z axis (direction of B) produce mirror field and the current components which are parallel to the z axis produce quadrupole field. Minimum-B in the anchor cells of the GAMMA10 tandem mirror is produced with the combination of baseball and racetrack coils and is connected with the axisymmetric mirror cells of circular flux tube through transition regions. Baseball coil at the transition region is called recircularizing baseball coil. Circular flux tube of the central cell mid-plane becomes highly elliptic at the transition regions of the anchor cell. At the fanning regions of the flux tube, the field lines gradually bulge in outward direction, i.e., field lines move towards the coil. However, ellipticity of the flux tube depends on the strength of the baseball coil current and also on the length of coil current in z direction. Therefore, calculation of the magnetic field lines is carried out when only the straight arms of the recircularizing baseball coils of GAMMA 10 are converted into arc shape as shown in Fig.19. Baseball coil currents are kept same as that for the present GAMMA 10 configuration. In is found that, the ellipticity of the flux tube at the outer transition region, at $z=644cm$, is improved by 34.8% (Fig.19(a)) and the stability of a field line is improved by 7% (Fig.19(b)) compared to the present GAMMA 10 magnetic field configuration. The improvement of stability is due to the reduction of bad curvature of the field lines in the anchor cell. The MHD stability of the magnetic field line, which crosses the central cell mid-plane at $(3.55, 3.55, 0)$ (unit in cm) is studied. The MHD stability of a magnetic field configuration is studied by calculating the line integral [85]

$$\Gamma(z) = \int_0^z \frac{1}{B}\left(\overline{P}_{||} + \overline{P}_\perp\right)\kappa_\varphi dz, \quad (2.4)$$

where, B is the magnetic field strength, $\overline{P}_{||}$ and $\overline{P}_\perp$ are the plasma pressures parallel and perpendicular to the magnetic field line, respectively and κ_φ is the normal curvature of the filed line. The MHD stable MFC requires $\Gamma(z) \geq 0$.

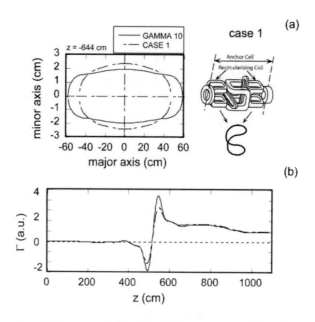

Figure 19. (a) Cross-section of the magnetic flux tube at the outer transition of anchor cell and modified shape of recircularizing base ball coil. (b) Running integral of flute-interchange mode of a field line which crosses the central cell mid-plane at $(3.55, 3.55, 0)$ (unit in cm).

2-7   Summary and Conclusions

Edge plasma at the non-axisymmetric MFC region of the anchor cell of the GAMMA 10 tandem mirror is studied using Langmuir probe, calorimeters, conducting plates (AP), H_α detectors, etc. Probe current asymmetry in minor axis direction of the elliptic flux tube at the outer transition region is found. The asymmetry increases in outward direction and is independent on experimental conditions. Mechanism of this asymmetry is explained with the plasma flow through the non-axisymmetric magnetic field regions, i.e., the direction of the shifting of plasma particles corresponds to that of ∇B and curvature drifts of ion. Since, due to the vast difference of parallel velocity and Larmor radius of passing ions and electrons, ions are significantly affected by the non-axisymmetric magnetic field configuration. The positive floating potentials of the conducting plates indicate that large amount of ions is drifting out compared to electrons. Moreover, it is found that the positive floating potentials of the plates are larger in the ion drift side. Similar asymmetry in the measurements by calorimeters are found, which indicates that heat flux is dominated by ion flux. Some impurity deposited areas are observed on the APs and the pattern of these areas can be explained with the mechanism of the desorption of impurities due to the interaction of the drifted out ions with the APs. Impurity deposited areas are in the opposite side of the ion drift due to ∇B and curvature drifts. This observation indicates that sputtered impurities from the plates of ion drift sides are redeposited on the plates at the opposite sides of ion drift. The evidence which strongly indicates the cold plasma formation in the anchor cell is obtained. Mechanism of the formation of cold plasma is explained from the viewpoint of interaction of the drifted out passing particles with the wall/grounded APs. Comparatively low temperature plasma can be formed due to the ionization of desorped neutrals, charge exchange process, etc. Significant effect of the floated APs on the GAMMA 10 plasma parameters is observed and is thought to be due to the reduction of plasma wall interaction in the way of self biasing effect of the floated APs. Using three dimensional Monte-Carlo code DEGAS ver. 63, a modeling of neutral transport is successfully performed in the GAMMA 10 anchor cell with non-axisymmetric configuration of the magnetic field, and the behavior of neutrals in this region is investigated on the basis of the experimental data of H_α line-emission. The computational result well predicts the experimental results, which shows that the hydrogen recycling and particle penetration in the anchor transition regions play an important roll on the behavior of neutral particles in this region. Ambient neutral pressure control experiment is performed using gas fueling rate modulation and significant improvement of anchor plasma parameters is obtained. This experiment suggests that enhanced ambient neutral pressure in the anchor cell has adverse effect to form high β plasma. Calculation of the magnetic field lines are carried out to optimize the non-axisymmetric magnetic field configuration of the anchor cell of GAMMA 10. It is found that the ellipticity of the flux tube at the transition region can be improved significantly only by changing the shape of recircularizing baseball coils. In this case, MHD stability of the configuration is also improved due to the reduction of bad curvature of the field lines in the anchor cell.

Non-axisymmetric magnetic field configuration in the anchor cell can introduce strong wall plasma interaction. Presence of enhance neutral pressure in the minimum-B region can produce higher density plasma [cf. Fig. 10(a)], i.e., higher potential plasma compared to the central cell plasma (as discussed in 2.3.4). Electrostatic reflection of the passing ions from the central cell by the potential of the anchor cell can introduce $\pi/2$ non-axisymmetric resonant diffusion. Resonant diffusion of the passing ion can limit the increment of the plasma density of the central cell [75]. In addition, the presence of desorped neutrals/impurity has adverse effect on the formation and sustaining of plug/barrier potentials in the plug/barrier cells [29,30] which are next to the anchor cell. Hence, it is important to care the design of the coils and their arrangement for the optimization of the non-axisymmetric magnetic field configuration in the anchor cell. Present investigations will help to improve the design of the tandem mirror with minimum-B anchor cell. Moreover, the experimental investigation in the GAMMA 10 plasma clarifies some limitations in the experiment, such as, density saturation, related problems on MHD stability, and plug/barrier potential formation.

3. PART B: THEORETICAL STUDY ON PLASMAS IN DIVERTOR ENVIRONMENT

Dust grains in the range of $10nm$ to $100\,\mu m$ can be immersed both in natural and laboratory plasmas. They can not remain neutral in plasma. Due to the interaction of electrons, ions, and background radiation, dust grains can be charged with the order of $Q_d \sim 10^3 e - 10^4 e$. The dust grains can be charged by absorbing electrons and ions flowing onto their surface. Due to the high mobility of electrons, the grain charge is usually negative. On the other hand, in a radiation background, dust grains which emit photoelectrons, may become positively charged. Negative and positive dust grains can coexist in some situations of dusty plasma [86,87]. In some special circumstances of dusty plasma, charging due to thermonic emission and secondary emission can be significant. In general, the presence of charged dust grains in plasmas can be characterized with respect to space scales by the following two conditions: 1) $a_d \ll \lambda_d < d$, and 2) $a_d \ll d < \lambda_d$, where a_d is the size of the grains, d is the average distance between any two grains, and λ_d is the plasma Debye length. In the first case, dust appears as isolated screened grains and the plasma is then called "dust-in-plasma". In the second case, the charged dust dynamics plays a vital role and the plasma becomes a true "dusty plasma". In relation to the masses of dust grains, dusty plasmas may be viewed in the two approaches. In the first, the random distribution of the massive, immobile, and highly charged particles can change the dispersion properties leading to new waves due to the motion of plasma particles around the correlated dust grains. In the other limit, dust can be taken as a third component of plasma having constant charge and mass. In this case, collective effects due to the motion of the massive and charged grains become evident and significant, leading to new low frequency wave modes.

Presently, theoretical and experimental researches on both dust-in-plasma and dusty plasma are proceeding at a rapidly increasing pace. Theoretical and experimental researches in dusty plasma have expanded into a wider range of problems, including studies of collective processes, i.e., waves and instabilities. Presence of relatively highly charged and massive dust grains in a plasma can modify or influence the collective phenomena of the plasma [88]. Experiments in the laboratory have verified the theoretical predictions of the electrostatic dust-acoustic wave [89,90], the electrostatic dust-ion-acoustic (DIA) wave [91,92], and the dust-lattice (DL) wave [93]. Magnetized dusty plasmas obviously support more additional electrostatic low frequency waves, such as dust-lower-hybrid wave having frequency below the ion-cyclotron frequency [94]. In high density low temperature plasma, the DLH mode arises because of the hybrid motion of the unmagnetized dust grains and the magnetized lighter plasma particles. If the electrons are considered as hot Boltzmann gas at temperature T_e in low density magnetized dusty plasma then DA mode is obtained, which propagates nearly perpendicular to the external static magnetic field, i.e., in the direction of DLH mode [95,96]. In this part, charging of dust grains in the case of unmagnetized plasma is given in section 3-1. Results of the theoretical study on the properties and instabilities of dust modes, such as dust-lower-hybrid and dust-acoustic modes in a divertor plasma environment are given in section 3-2.

3-1   Charging of Dust Grain

To derive expressions for charged particle currents (I) to dust grain as a function of dust surface potential, ϕ_d, we have considered a spherical dust grain immerged in plasma. In this derivation, we have taken the advantages of the procedure in the deduction of electric probe formulas proposed by Medicus [97] and moreover, have followed the book in Ref. 98. The current to the dust grain for the particles with uniform initial velocities v is given by

$$I = \pi q \int p_g^2 v f(v) dv, \qquad (3.1)$$

where πp_g^2 is the "collision" cross-section of the spherical dust grain for particles with the velocities v. p_g is the impact parameter of a particle which grazes the dust grain and $f(v)$ is the velocity distribution of the particles.

The impact parameter p_g for which the particles reach the dust grain is given by [98]

$$p_g^2 = a_d^2 \left(1 - \frac{2q\phi_d}{mv^2} \right). \qquad (3.2)$$

In retarding field, i.e., when $q\phi_d > 0$, the particles with $p_g = a_d$ reach the dust grain, provided their velocity in the following range

$$v \geq \left(\frac{2q\phi_d}{m} \right)^{\frac{1}{2}}. \qquad (3.3)$$

Using the velocity limit [Eq.(3.3)] and the value of p_g [Eq. (3.2)], we get the current to dust grain in retarding field from Eq. (3.1)

$$I_r = \pi a_d^2 q \int_{\left(\frac{2q\phi_d}{m} \right)^{\frac{1}{2}}}^{\infty} \left(1 - \frac{2q\phi_d}{mv^2} \right) v f(v) dv. \qquad (3.4)$$

Dust charging current integral in retarding field - Eq. (3.4) - becomes the current integral in accelerating field ($q\phi_d < 0$) for thick sheath when the limit of velocity integration is extended from 0 to ∞ [99]. The motion of the particles to the dust grain is said to be "orbit limited" (OL). In OL current, most of the particles entering the sheath miss the dust grain, only a few hit it. The OL current to dust grain in accelerating field is thus

$$I_a = \pi a_d^2 q \int_0^\infty \left(1 - \frac{2q\phi_d}{mv^2}\right) v f(v) dv . \qquad (3.5)$$

We now apply the results [Eqs. (3.4) and (3.5)] to Maxwellian distribution of particles of density n_o, where

$$f(v) = 4\pi n_o \left(\frac{m}{2\pi k_B T}\right)^{\frac{3}{2}} v^2 \exp\left(-\frac{mv^2}{2k_B T}\right). \qquad (3.6)$$

The straight forward integration of Eqs.(3.4) and (3.5) yields the charged particles current to dust grain in retarding and accelerating fields, respectively

$$I_r = \pi a_d^2 q n_o \left(\frac{8k_B T}{\pi m}\right)^{\frac{1}{2}} \exp\left(-\frac{q\phi_d}{k_B T}\right), \qquad (3.7)$$

and

$$I_a = \pi a_d^2 q n_o \left(\frac{8k_B T}{\pi m}\right)^{\frac{1}{2}} \left(1 - \frac{q\phi_d}{k_B T}\right). \qquad (3.8)$$

If the particles stream with an arbitrary streaming velocity (v_o), then the dust charging current expressions are much more complicated [99]. However, in the limit $v_o \gg C$ (where C is the thermal velocity of the plasma particles) and in the case of accelerating field $q\phi_d < 0$, we can write the approximate expression for streaming particle current to dust grain,

$$I_{as} = \pi a_d^2 q n_o v_o \left(1 - \frac{2q\phi_d}{mv_o^2}\right), \qquad (3.9)$$

which may be readily obtained by integrating over δ -function distributions. For both signs of charge (q), the quantity $q\phi_d$ is positive for retarding and negative for accelerating fields.

3-2   Properties and Excitation of Low Frequency Dust Modes

To derive the dispersion relation of low frequency electrostatic dust modes below the ion cyclotron frequency in a streaming magnetized dusty plasma with dust charge fluctuation following two assumptions are considered. At first, we compare the dust grain size with the Larmor radii of the plasma particles. The expression for electron and ion Larmor radii are $r_e = 3.37 \times 10^{-6} \sqrt{T_e}/B$ and $r_i = 1.44 \times 10^{-4} \sqrt{T_i}/B$, respectively, where $T_{e,i}$ are the specific particle temperature. Typically, $B = \sim 3T$ in the divertor region and $B = \sim 0.05T$ in the end plate region of a mirror machine, and the size of the dust grains is in the range of $10nm$ to $100 \mu m$ for microparticles. Edge plasma parameters are significantly controlled by the core plasma parameters, which depend on the experimental conditions. However, in the case of divertor plasma when $T_i = 7eV$, $T_e = 20eV$, and $B = 0.5T$, then Larmor radii of the particles become: $r_i = 7.6 \times 10^{-1} mm$ and $r_e = 3.0 \times 10^{-2} mm$. Therefore, the dust grain size (a_d) can be much small compared to the average Larmor radius (r_j) of the plasma particles, i.e., $r_j >> a_d$, where j stands for the charged plasma species. In this limit, the charging equation of dust grain in unmagnetized plasma, Eqs. (3.7)-(3.9), can be applied to the magnetized plasma. Secondly, in a divertor region, plasmas in scrap-off layer (SOL) flow at the velocity of sound along the lines of magnetic force just outside the separatrix into the divertor plates. The sheath formation close to the divertor plate gives plasma potential variation in the boundary. We consider the streaming of ions (v_{io}) relative to the dust grains in the direction of B and v_{io} is larger than the ion thermal speed (C_i). v_{io} in the plasma-sheath interface region may be of the order of ion sound speed $C_s = \sqrt{2k_B(T_e + T_i)/m_i}$, which is larger than the ion thermal speed $C_i = \sqrt{2k_B T_i/m_i}$ in divertor plasma where $T_e/T_i \approx 3$. The quantities T_j and m_j are temperature and mass of the species j, respectively and k_B is Boltzmann constant. However, in the limit $v_{io} >> C_i$, we can use the Eq. (3.9), the approximate expression for streaming particle current to dust grain in the accelerating field.

3.2.1 Fluid Equations

We have considered a homogeneous and uniform magnetized dusty plasma consists of electrons, ions, negatively charged dust grains, and neutrals. The external magnetic field (B) is in the z-direction. The neutral gas is taken to be at rest. The dust grain charge is considered as negative, i.e., $Q_d = - Z_d e$, where Z_d is the number of charge on a dust grain and e is the electron charge. Let us consider a low frequency electrostatic wave propagating obliquely ($K_x^2 >> K_z^2$) to the external magnetic field with propagation vector (K) lying in the xz-plane. Due to the presence of this mode (ω, K), the dust grain will acquire a perturbed charge, Q_{d1}.

Continuity and momentum equations for different species of the dusty plasma are taken as follows, respectively:

$$\frac{\partial n_j}{\partial t} + \nabla \cdot (n_j \mathbf{v}_j) = 0, \qquad (3.10)$$

and

$$n_{d,i} m_{d,i} \frac{\partial \mathbf{v}_{d,i}}{\partial t} + n_{d,i} m_{d,i} \mathbf{v}_{d,i} \cdot \nabla \mathbf{v}_{d,i} + k_B T_{d,i} \nabla n_{d,i}$$
$$- Q_{d,i} n_{d,i} \mathbf{v}_{d,i} \times B = -\nu_{d,i} n_{d,i} m_{d,i} \mathbf{v}_{d,i}$$

(3.11)

$$k_B T_e \nabla n_e + e n_e \mathbf{v}_e \times B = -\nu_e n_e m_e \mathbf{v}_e, \qquad (3.12)$$

where $j = d, i, e$ and d, i, e denote respectively, the dust grain, ion, and electron quantities. Moreover, n_j, v_j, and Q_j are the density, velocity, and charge of the species j and v_j is the collisional frequency of neutrals with charged species j.

The continuity and momentum equations (3.10) to (3.12) will be coupled to the following set of equations, namely:
Poisson's equation

$$- \varepsilon_o \nabla^2 \phi = e[n_i - Z_d n_d - n_e], \qquad (3.13)$$

quasi-neutrality equation

$$n_i = n_e + Z_d n_d, \qquad (3.14)$$

and basic dust charging equation

$$\frac{dQ_d}{dt} = I_e + I_i, \qquad (3.15)$$

where I_e and I_i are, respectively, the electron and ion currents collected by the dust grains. Moreover, ε_o and ϕ present the electric permittivity of vacuum and electric potential, respectively. The electron and ion currents involved in the basic dust charging equation (3.15) can be written as:

$$I_e = -e n_e \pi a_d^2 \left(\frac{8 k_B T_e}{\pi m_e} \right)^{\frac{1}{2}} \exp\left[\frac{e(\phi_p - \phi_d)}{k_B T_e} \right]$$

$$I_i = (e n_i v_{io}) \pi a_d^2 \left[1 - \frac{2e(\phi_p - \phi_d)}{m_i v_{io}^2} \right], \qquad (3.16)$$

where a_d, ϕ_p, and ϕ_d are the dust radius, bulk plasma potential, and dust grain surface potential, respectively.

Linearized Eqs. (3.10)-(3.16) by the usual technique and the first-order quantities are assumed to vary in space and time as $\exp[i(K_x X + K_z Z - \omega t)]$. We get the perturbed quantities as follows [100,101]:

$$n_{i1} = \frac{e n_{io} R \phi_1}{m_i \left(\Omega_i - c_i^2 R \right)}, \qquad (3.17)$$

$$n_{e1} = \frac{i e n_{eo} F \phi_1}{m_e \left(\omega + i c_e^2 F \right)}, \qquad (3.18)$$

$$n_{d1} = - \frac{e Z_{do} n_{do} L \phi_1}{m_d \left(\omega - c_d^2 L \right)}, \qquad (3.19)$$

$$\phi_1 = \frac{e}{\varepsilon_o K^2} \left(n_{i1} - n_{e1} + \frac{Q_{do}}{e} n_{d1} + \frac{n_{do}}{e} Q_{d1} \right), \qquad (3.20)$$

and
$$Q_{d1} = \frac{i|I_{eo}|}{\omega + i \eta} \left[\frac{n_{i1}}{n_{io}} - \frac{n_{e1}}{n_{eo}} \right], \qquad (3.21)$$

where

$$R = \frac{\left(K_x^2 + K_z^2 \right)\left(\Omega_i + i v_i \right)^2 - K_z^2 \omega_{ci}^2}{\left[\left(\Omega_i + i v_i \right)^2 - \omega_{ci}^2 \right]\left(\Omega_i + i v_i \right)},$$

$$F = \frac{\left(K_x^2 + K_z^2 \right) v_e^2 + K_z^2 \omega_{ce}^2}{\left(v_e^2 + \omega_{ce}^2 \right) v_e}, \qquad (3.22)$$

$$L = \frac{\left(K_x^2 + K_z^2 \right)\left(\omega + i v_d \right)^2 - K_z^2 \omega_{cd}^2}{\left[\left(\omega + i v_d \right)^2 - \omega_{cd}^2 \right]\left(\omega + i v_d \right)},$$

$$\eta = e\left(|I_{eo}|/C \right)\left(1/k_B T_e + 1/w_o \right), \quad and \quad w_o = m_i v_{io}^2/2 - e Q_{do}/C.$$

The fluctuated charge of dust will decay in the rate η and C is the dust capacitance. The Doppler shifted frequency, $\Omega_i = \omega - K_z v_{io}$ and cyclotron frequency of the species, j, $\omega_{cj} = Q_j B/m_j$. $|I_{eo}|$ is the magnitude of equilibrium electron current flowing into the dust grain and at equilibrium $I_{eo} = I_{io}$ is considered.

Using the value of ϱ_{d1} from equation (3.21), we get the linearized Poission's equation (3.20) as:

$$\phi_1 = \frac{e}{\varepsilon_o K^2}\left[n_{i1}\left\{1 + \frac{i\beta}{\omega + i\eta}\frac{n_{eo}}{n_{+o}}\right\} - n_{e1}\left\{1 + \frac{i\beta}{\omega + i\eta}\right\} - Z_{do}n_{d1}\right],$$

(3.23)

where $\beta = \left(|I_{eo}|/e\right)\left(n_{do}/n_{eo}\right) = 10^{-1}\pi a_d^2 n_{do}C_e$ and the $i\beta$ terms are arising through coupling to dust charge fluctuations which result as a response to collective plasma perturbation. The dispersion relation for the dust modes is then obtained from Eq. (3.23) by putting the value of n_{i1}, n_{e1}, and n_{d1} from Eqs. (3.17), (3.18), and (3.19), respectively. The dispersion relation is then given by:

$$1 = \frac{\omega_{pi}^2 R}{K^2\left(\Omega_i - C_i^2 R\right)}\left\{1 + \frac{i\beta}{\omega + i\eta}\frac{n_{eo}}{n_{io}}\right\} - \frac{i\omega_{pe}^2 F}{K^2\left(\omega + iC_e^2 F\right)}\left\{1 + \frac{i\beta}{\omega + i\eta}\right\} + \frac{\omega_{pd}^2 L}{K^2\left(\omega - c_{dL}^2\right)},$$

(3.24)

where the plasma frequency of the species, j, is $\omega_{pj}^2 = n_j Q_j{}^2/\varepsilon_o m_j$. The dispersion relation of Eq. (3.24) for the dust modes in collisional, streaming magnetized dusty plasmas with dust charge fluctuation is the basic one for the analysis of low frequency dust modes.

3.2.2 Low Frequency Dust Modes

As in the DA wave, we analogously consider the presence of a low frequency electrostatic dust mode below the ion-cyclotron frequency propagating nearly perpendicular ($K_x^2 \gg K_z^2$) to the external static magnetic field in the magnetized dusty plasma where the dynamics of the dust particles may play an important role. In this case, we assume that the dust mode under consideration satisfies the following conditions:

$$\omega_{cd} \ll \omega \ll \omega_{ci} \ll \omega_{ce}; \qquad K_z C_d, K_z C_i \ll \omega \ll K_z C_e.$$

Under these conditions, the highly charged and massive dust grains can be taken to be cold ($C_d = 0$) and unmagnetized ($\omega_{cd} = 0$). Ions are cold ($C_i = 0$), but strongly magnetized. Electrons form a hot Boltzmann gas at temperature T_e. As $m_d \gg m_i$, m_e, collision of dust grains with neutrals may be neglected ($\nu_d = 0$). Under these conditions Eq. (3.22) becomes:

$$R = -\frac{K_x^2\left(\Omega_i + i\nu_i\right)}{\omega_{ci}^2} + \frac{K_z^2}{\left(\Omega_i + i\nu_i\right)}, \quad F = \frac{K_x^2 \nu_e}{\omega_{ce}^2} + \frac{K_z^2}{\nu_e}, \quad L = \frac{K^2}{\omega},$$

and the Eq. (3.24) then gives the dispersion relation for our required dust modes:

$$\left(\omega^2 - \omega_{DM}^2\right)\left(\omega + i\eta + i\beta' + i\nu_i' + i\nu_e'\right) = -i\omega_{DM}^2\left(\beta' + \nu_i' + \nu_e'\right), \qquad (3.25)$$

where

$$\omega_{DM}^2 = \frac{\dfrac{K_z^2 \omega_{pi}^2}{K^2 \Omega_i^2} \omega^2 + \omega_{pd}^2}{1 + \dfrac{1}{K^2 \lambda_{De}^2} + \dfrac{K_x^2 \omega_{pi}^2}{K^2 \omega_{ci}^2}}, \qquad (3.26)$$

$$\beta' = \beta \frac{\dfrac{K_x^2 \omega_{pi}^2 n_{eo}}{K^2 \omega_{ci}^2 n_{io}} + \dfrac{1}{K^2 \lambda_{De}^2}}{1 + \dfrac{1}{K^2 \lambda_{De}^2} + \dfrac{K_x^2 \omega_{pi}^2}{K^2 \omega_{ci}^2}}, \qquad (3.27)$$

$$\nu_i' = \nu_i \frac{\dfrac{K_x^2 \omega_{pi}^2 \omega}{K^2 \omega_{ci}^2 \Omega_i}}{1 + \dfrac{1}{K^2 \lambda_{De}^2} + \dfrac{K_x^2 \omega_{pi}^2}{K^2 \omega_{ci}^2}}, \qquad (3.28)$$

and

$$\nu_e' = \nu_e \frac{\dfrac{\omega^2}{K_z^2 K^2 \lambda_{De}^2 C_e^2}}{1 + \dfrac{1}{K^2 \lambda_{De}^2} + \dfrac{K_x^2 \omega_{pi}^2}{K^2 \omega_{ci}^2}}. \qquad (3.29)$$

The dispersion relation (3.25) is a cubic equation in ω. The first bracket describes the two electrostatic dust modes, the second bracket indicates the damped low frequency mode related with dust charge fluctuation and collisional terms and the right hand side can be interpreted as a coupling term. A perturbative solution of Eq. (3.25) [where η and $(\beta' + \nu_i' + \nu_e')$ are very small] gives three roots:

$$\omega = \mp \omega_{DM} - \frac{i}{2} (\beta' + \nu_i' + \nu_e'), \qquad (3.30)$$

$$\omega = -i\eta \left[1 + \frac{\eta}{\omega_{DM}^2} (\beta' + \nu_i' + \nu_e') \right]. \qquad (3.31)$$

Eq. (3.31) gives a low frequency damped mode. From Eq. (3.30), it is seen that the dust charge fluctuation, ion and electron collisional terms are linearly added to the mode. This dispersion relation will be used for the analysis of the properties and instabilities of the dust modes below the ion cyclotron frequency in the following sections.

3.2.3 Two-Stream Instability

To discuss the two-stream instability of low frequency dust modes, we consider $\beta' = v'_i = v'_e = 0$. In this case Eq. (3.30) becomes:

$$\omega^4 + M\omega^3 + N\omega^2 + O\omega + P = 0, \quad (3.32)$$

where

$$M = -2K_z v_{io}, \ N = K_z^2 v_{io}^2 - \omega_{DM}'^2, \ O = 2\frac{\omega_{pd}^2}{A} K_z v_{io}, \ P = -\frac{\omega_{pd}^2}{A}(K_z v_{io})^2, \text{, and}$$

$$\omega^2 = \left(1 + K_z^2 \omega_{pi}^2 / K^2 \omega_{pd}^2\right)\omega_{pd}^2 / A.$$

If $K_z v_{io} \gg \omega'_{DM}$, then $N\omega^2 \gg |M|\omega^3$ and $|P| \gg O\omega$. Thus, equation (3.32) reduces to quadratic equation as follows:

$$\omega^4 + N\omega^2 + P = 0. \quad (3.33)$$

Two roots of the equation (3.33) are as:

$$\omega_{1,2}^2 = -\frac{1}{2} N \pm \left(\frac{1}{4} N^2 - P\right)^{1/2}, \quad (3.34)$$

The root ω_1^2 is positive real quantities and therefore, there can be no temporal growth or decay of the amplitude of dust modes. On the other hand, ω_2^2 is a negative real quantity. Therefore, ω_2 has two imaginary values (one positive and one negative). The positive imaginary value of ω_2 corresponds to an unstable mode and the growth rate of the mode is given by

$$\gamma_2 = \left[\frac{1}{2} N + \left(\frac{1}{4} N^2 - P\right)^{1/2}\right]^{1/2}$$

$$= \left[\frac{1}{2}(K_z^2 v_{io}^2 - \omega_{DM}'^2) + \left(\frac{1}{4}(K_z^2 v_{io}^2 - \omega_{DM}'^2)^2 + \frac{K_z^2 v_{io}^2 \omega_{pd}^2}{1 + \frac{1}{K^2 \lambda_{De}^2} + \frac{K_x^2 \omega_{pi}^2}{K^2 \omega_{ci}^2}}\right)^{1/2}\right]^{1/2 \cdot}$$

(3.35)

On the other hand, if $\omega'_{DM} \gg K_z v_{io}$, then N becomes negative and $|N|\omega^2 \gg |P|, O\omega, |M|\omega^3$. Thus, equation (3.32) reduces to $\omega^2 = N$, and there is no instability. Thus, the two-stream instability, of the electrostatic dust modes below the ion cyclotron frequency in streaming dusty plasmas which propagate nearly perpendicular to the external magnetic field, arises only when the ion streaming velocity becomes much larger than the wave parallel phase velocity.

3.2.4 Properties of DLH and DA Modes

For simplicity of the analysis of low frequency, low phase velocity electrostatic dust modes in the magnetized dusty plasma, two important cases of dusty plasma are considered: Case A: high density dusty plasma and Case B: low density dusty plasma.

Case A: High density dusty plasma, i.e., $\omega_{pi}^2 / \omega_{ci}^2 > 1/K^2 \lambda_{De}^2$

For high density plasma, the dispersion relation of the DLH mode is obtained from Eq. (3.30):

$$\omega = \omega_{DLH} \left[1 + \frac{K_z^2}{K^2} \frac{\omega_{pi}^2 \omega^2}{\omega_{pd}^2 \Omega_i^2} \right]^{1/2} - \frac{i}{2} \beta \frac{n_{eo}}{n_{io}} - \frac{i}{2} \nu_i \frac{\omega}{\Omega_i} - \frac{i}{2} \nu_e \frac{\omega^2}{K_z^2 c_e^2} \frac{1/\left(K^2 \lambda_{De}^2\right)}{\omega_{pi}^2 / \omega_{ci}^2}, \qquad (3.36)$$

where DLH frequency, $\omega_{DLH} = \omega_{pd} \omega_{ci} / \omega_{pi} = \sqrt{\omega_{ci} \omega_{cd} \left(Z_{do} n_{do} / n_{io}\right)}$.

Case B: Low density dusty plasma, i.e., $1/K^2 \lambda_{De}^2 > \omega_{pi}^2 / \omega_{ci}^2$

For low density plasma, dispersion relation of DA mode is obtained from Eq. (3.30):

$$\omega = KC_D \left[1 + \frac{K_z^2}{K^2} \frac{\omega_{pi}^2 \omega^2}{\omega_{pd}^2 \Omega_i^2} \right]^{1/2} - \frac{i}{2} \beta - \frac{i}{2} \nu_i \frac{\omega}{\Omega_i} \frac{\omega_{pi}^2 / \omega_{ci}^2}{1/\left(K^2 \lambda_{De}^2\right)} - \frac{i}{2} \nu_e \frac{\omega^2}{K_z^2 c_e^2},$$

(3.37)

where KC_D is DA frequency and $C_D = C_e \omega_{pd} / 2\omega_{pe}$.

The natural DLH and DA modes can be obtained from the equations (3.36) and (3.37) respectively, when $K_z V_{io} = \beta = \nu_i = \nu_e = 0$. The second and fourth terms of the right hand side (RHS) of Eqs. (3.36) and (3.37) present the dust charge fluctuation and electron neutral collisional effects, respectively, of the respective mode, which gives the damping of the mode. The third terms of the RHS of these equations present the ion neutral collisional effect. From the third terms, it is seen that if ions streaming velocity exceeds the wave parallel phase velocity, then the collisions of ions with neutral gas molecules give destabilizing effect provided the other damping effects are negligible. Growth rate of two-stream instability of the DLH and DA modes can be obtained easily from the Eq. (3.35) using the conditions of high and low density plasmas, respectively.

3-3   Discussions and Conclusions

In high density low temperature magnetized dusty plasma, the DLH mode arises due to the dynamics of negatively charged unmagnetized dust grains and the magnetized ions [cf. Eq. (3.36)]. If the electrons are considered as hot Boltzmann gas at temperature T_e in low density magnetized dusty plasma, then DA mode is obtained, which propagating nearly perpendicular to the external static magnetic field [cf. Eq. (3.37)]. It is found that the ions streaming velocity is coupled directly with the DLH and DA modes. In the absence of dust charge fluctuation and collisional effects, two-stream instability of these modes occurs when the ion streaming velocity becomes much larger than the parallel phase velocity of the respective mode. Growth rate of two-stream instability of the DLH and DA modes can be obtained easily from the Eq. (3.35) using the conditions of high and low density plasmas, respectively. The dust charge fluctuation effect linearly added to the collisional effect in any case of the modes. Dust charge fluctuation and electron neutral collisions give the damping of either of these modes. Since $K_z C_e \gg \omega$ for any of the modes and $\omega_{pi}^2 / \omega_{ci}^2 > 1 / K^2 \lambda_{De}^2$ in the case of DLH mode, the electron neutral collision effect is negligible for both the modes [cf. Eqs. (3.36) and (3.37)]. It is also found that the streaming energy of ions is coupled with the DLH and DA modes through the collisions of ions with the neutrals. If ions streaming velocity exceeds the wave parallel phase velocity, then the collision of ions with neutral gas atoms/molecules gives destabilizing effect on the dust modes provided the effects of dust charge fluctuation and electron neutral collision are negligible.

The present investigation should be useful in understanding the various properties and effects of extremely low frequency dust modes in the divertor plasma.

REFERENCES

[1]   F. Wagner, M. Keilhacker, *J. Nucl. Mater.* 121, 103 (1984).
[2]   M. Keilhacker, *Plasma Phys. Controll. Fusion* 29, 1401 (1987).
[3]   K. H. Burrell, S. L. Allen, G. Bramson, *Plasma Phys. Controll. Fusion* 31, 1649 (1989).
[4]   Gibson, *Nucl. Fusion* 16, 546 (1976).
[5]   P. H. Rebut, B. Green, in *Plasma Physics and Controlled Nuclear Fusion Research 1976* (Proc. 6th Int. Conf. Berchtesgaden, 1976), vol.2, IAEA, Vienna (1977) 3.
[6]   J. Wesson, C. Gowers, W. Han, et al., in *Controlled Fusion and Plasma Physics* (Proc. 12th Eur. Conf. Budapest, 1985), vol. 9F, Part I, European Physical Society (1985) 147.
[7]   P. C. Stangeby, G. M. McCracken, *Nucl. Fusion* 30, 1225 (1990).
[8]   D. E. Post, R. Behrisch (Eds), *Physics of Plasma-Wall Interactions in Controlled Fusion*, Plenum Press, New York (1986).
[9]   Y. Nakashima, K. Yatsu, K. Islam, et al., *J. Plasma and Fusion Res.* 75, 1211 (1999).
[10]   M. K. Islam, et al., *J. Phy. Soc. Japan.* 69, 2493 (2000).
[11]   G.I. Dimov, V. V. Zakaidakov, and M. E. Kishinevskii, *Fiz. Plasmy* 2, 597 (1976) [English transl.: *Sov. J. Plasma Phys.* 2, 326-333].
[12]   T. K. Fowler and B. G. Logon, *Comments Plasma Phys. Controlled Fusion Res.* 2, 167 (1977).

[13] T. Tamano, et al., in *Plasma Physics and Controlled Nuclear Fusion Research* 1994 (Proc. 15th Int. Conf. Seville, 1994), vol.2, IAEA, Vienna (1995) 399.

[14] T. Cho, H. Higaki, M. Hirata, H. Hojo, et al., *20th IAEA Fusion Energy Conference*, EX/9-6Rd.

[15] J. Kesner, R. S. Post, B. D. McVey, and D. K. Smith, *Nucl. Fusion* 22, 549 (1982).

[16] N. Hershkowitz, R. A. Breun, D. Brouchous, in *Proc. 9th Int. Conf. Plasma Physics and Controlled Nuclear Fusion Research*, Baltimore, MD (1982), IAEA, Vienna, 1983, vol.1, p.553.

[17] T. C. Simonen, D. E. Baldwin, S. L. Allen, in *Proc. 9th Int. Conf. Plasma Physics and Controlled Nuclear Fusion Research*, Baltimore, MD (1982), IAEA, Vienna, 1983, vol.1, p.519.

[18] D. E. Baldwin, B. G. Logan, T. C. Simonen, *Physics Basis for MFTF-B*, Lawrence Livermore National Laboratory, Livermore, CA, UCID-18496 (1980).

[19] K. I. Thomassen, V. N. Karpenko, *An Axicell Design for the End Plugs of MFTF-B*, Lawrence Livermore National Laboratory, Livermore, CA, UCID-19318 (1982).

[20] M. Inutake, K. Ishii, A. Itakura, in *Proc. 9th Int. Conf. Plasma Physics and Controlled Nuclear Fusion Research*, Baltimore, MD (1982), IAEA, Vienna, 1983, vol.1, p.545.

[21] G. I. Dimov and G. V. Roslyakov, *A Trap with Ambipolar Plugs*, (Institute of Nuclear Physics, Novosibirsk, USSR, Preprint 80-1520), published by Lawrence Livermore National Laboratory, Livermore, CA, UCRL-TRANS-11670 (1981).

[22] T. D. Akhmetov, V. S. Belkin, I. O. Bespamyatnov, et al., *Trans. Fusion Tech.* 43, 58 (2003).

[23] A. Ivanov, G. F. Abdrashitov, A. V. Anikeev, et al., *Trans. Fusion Tech.* 43, 51 (2003).

[24] M. Kwon, J. G. Bak, K. Choh, et al., *Trans. Fusion Tech.* 43, 23 (2003).

[25] Y. Nakashima, K. Yatsu, et al., *J. Nucl. Mater.* 196 - 198, 493 (1992).

[26] M. K. Islam, M. Salimullah, K. Yatsu, et al., *Nucl. Fusion* 43, 914 (2003).

[27] S. K. Erents, J. A. Tagle, G. M. McCracken, et al, *Nucl. Fusion* 28, 1209 (1988).

[28] P. Bogen, H. Hartwig, E. Hintz, et al., *J. Nucl. Mater.* 128 & 129, 157 (1984).

[29] W. L. Hsu, *J. Nucl. Mater.* 128 & 129, 500 (1984).

[30] S. L. Allen, et al., *Plasma-Wall Interactions in Tandem Mirror Machines, in Physics of Plasma-Wall Interactions in Controlled Fusion*, Plenum Press, New York (1986), Editted by D. E. Post, R. Behrisch.

[31] Mase, A. Itakura, M. Inutake, et al., *Nucl. Fusion* 31, 1725 (1991).

[32] T. Saito, Y. Tatematsu, I. Katanuma, K. Yatsu, *J. Plasma Fusion Res.* 78, 591 (2002).

[33] M. K. Islam, Y. Nakashima, T. Natori, et al., *Trans. Fusion Tech.* 43, 177 (2003).

[34] S. L. Allen, et al., *Nucl. Fusion* 27, 2139 (1987).

[35] D. Reiter, *et al.*, *Plasma Phys. Contrib. Fus.* 33, 1579 (1991).

[36] D. P. Stotler, *et al.*, *Contrib. Plasma Phys.* 34, 392 (1994).

[37] D. Herfetz, D. Post, M. Petravic et al., *J. Comput. Phys.* 46, 309 (1982).

[38] Y. Nakashima, *et al.*, *J. Nucl. Mater.* 241-243, 1011 (1997).

[39] S. Kobayashi, *et al.*, *J. Nucl. Mater.* 266-269, 566 (1999).

[40] Y. Nakashima, *et al.*, *J. Nucl. Mater.* 313-316, 553 (2003).

[41] G. M. McCracken, P. E. Stott, *Nucl. Fusion* 19, 889 (1979).

[42] Koma (Ed.), Desorption and Related Phenomena Relevent to Fusion Devices, *Rep. IPPJ-AM-22*, Inst. of Plasma Physics, Nagoya Univ. (1982).

[43] G. M. McCracken, D. H. J. Goodall, *Nucl. Fusion* 18, 537 (1978).

[44] D. H. J. Goodall, et al., *J. Nucl. Mater.* 76 & 77, 492 (1978).

[45] J. Roth, *J. Nucl. Mater.* 176 & 177, 132 (1990).

[46] G. M. McCracken, *J. Nucl. Mater.* 93 & 94, 3 (1980).

[47] M. Rubel, *Nucl. Fusion* 41, 1087 (2001).

[48] J. Winter, et al., *J. Nucl. Mater.* 162-164, 713 (1989).

[49] U. Samm, et al., *J. Nucl. Mater.* 222-224, 25 (1995).

[50] N. Noda, V. Philipps, R. Neu, *J. Nucl. Mater.* 241-243, 27 (1997).

[51] V. Philipps, et al., *J. Nucl. Mater.* 258-263, (1998).

[52] M. Rubel, *J. Nucl. Mater.* 283-287, 1089 (2000).

[53] V. Philipps, et al., *Plasma Phys. Control. Fusion* 42, B293, (2000).

[54] L. Keller, et al., *J. Nucl. Mater.* 111 & 112, 493 (1982).

[55] K. Narihara, K. Toi, Y. Hamada, et al., *Nucl. Fusion* 37, 1177 (1997).

[56] J. Winter, *Plasma Phys. Control. Fusion* 40, 1201, (1998).

[57] J. Wesson, *Tokamaks*, Clarendon Press, Oxford (1987).

[58] W. M. Stacey Jr., *Fusion Plasma Analysis*, Wiley, New York (1981).

[59] Yu. Chutov, O. Kravchenko, P. Schram, V. Yakovetsky, *Physica B* 262, 415 (1999).

[60] Yu. I. Chutov, O. Yu. Kravchenko, P. P. J. M. Schram, *Physica B* 128, 11 (1996).

[61] K. Yatsu, L. G. Bruskin, T. Cho, M. Hamada, M. Hirata, et al., *Trans. Fusion Tech.* 35, 52 (1999); K. Yatsu, et al., *Nucl. Fusion* 41, 613 (2001).

[62] S. Miyoshi, T. Cho, H. Hojo, M. Ichimura, K. Ishii, et al., *Plasma Physics and Controlled Nuclear Fusion Research* (IAEA, Vienna, 1991) vol.2, p.539.

[63] M. Inutake, T. Cho, M. Ichimura, K. Ishii, A. Itakura, et al., *Phys. Rev. Lett.* 55, 159 (1985).

[64] M. Ichimura, M. Inutake, S. Adachi, D. Sato, F. Tsuboi, et al., *Nucl. Fusion* 28, 799 (1988).

[65] T. Cho, M. Ichimura, M. Inutake, K. Ishii, A. Itakura, et al.: *Plasma Physics and Controlled Nuclear Fusion Research* (IAEA, Vienna, 1985) vol.2, p. 275.

[66] K. Yatsu et al., *Proc. APFA 1998 and APPTC 1998.*

[67] M. K. Islam, *Ph.D. Thesis*, University of Tsukuba, Japan, (1999).

[68] Y. Nakashima, et al., *J. Nucl. Mater.* 290-293, 683 (2001).

[69] Y. Nakashima, M. K. Islam, T. Natori, et al., *Rev. Sci. Instrument* 75, 4308 (2004).

[70] Yu. V. Gott, M. S. Ioffe, et al., *Nucl. Fusion Suppl.* part 3, 1045 (1962).

[71] R. W. Moir and R. F. Post, *Nucl. Fusion* 9, 253 (1969).

[72] T. C. Simonen, S. L. Allen, D. E. Baldwin, T. A. Casper, J . F. Clauser, et al., *Nucl. Fusion Suppl.* 2, 255 (1985).

[73] R. S. Post, M. Gerver, and J. Kesner, *Nucl. Fusion Suppl.* 2, 285 (1985).

[74] M. Inutake, S. Furukawa, S. Tanaka, R. Katsumata, A. Ishihara, et al., *Trans. of Fusion Tech.* 27, 409 (1995).

[75] D. D. Ryutov and G. V. Stupakov, *JETP Lett.* 26, 174 (1977); *Fiz. Plasmy* 4, 501 (1978); *Dokl. Akad. Nauk. SSSR* 240, 1086 (1978).

[76] F. F. Chen. *Introduction to Plasma Physics and Controlled Fusion*, 2nd ed., Vol. 1, Plenum Press, New York and London.

[77] R. F. Post, *Nucl. Fusion* 27, 1577 (1987).

[78] K. Ikeda, Y. Nagayama, T. Aota, et al., *Phys. Rev. Lett.* 78, 3872 (1997).

[79] W. M. Nevis and L. D. Pearlstein, *Phys. Fluids* 31, 1988 (1988).

[80] N. Yamaguchi, *et al.*, *Proc. 15th European Conf. on Controlled Fusion and Plasma Heating* (Dubrovnik, 16-20 May, 1988) Vol.12B, part-II, p.593.

[81] M. Yoshikawa, *et al.*, *Trans. Fusion Tech.* 35, 273 (1999).

[82] L. C. Johnson, E. Hinnov, *J. Quant. Spectrosc. Radiat. Transfer.* 13, 333 (1973).

[83] R. K. Janev, W. D. Langer, K. Evans Jr., D. E. Post Jr., *Elementary Processes in Hydrogen-Helium Plasmas*, Springer, Berlin, 1987.

[84] Y. Nakashima, et al., *Rev. Sci. Instrum.* 74, 2115 (2003).

[85] L. S. Hall and B. McNamara, *Phys. Fluids* 18, 552 (1975).

[86] K. Goertz, *Rev. Geophys.* 27, 271 (1989).

[87] M. Horanyi, B. Walch, S. Robertson, and D. Alexander, *J. Geophys. Res.* 103, 8575, (1998).

[88] P. K. Shukla, *Phys. Plasmas* 8, 1791 (2001).

[89] J. B. Piper and J. Goree, *Phys. Rev. Lett.* 77, 3137 (1996).

[90] Barken, N. D'Angelo, and R. L. Merlino, *Planet. Space Sci.* 44, 239 (1996).

[91] Y. Nakamura, H. Bailumg, and P. K. Shukla, *Phys. Rev. Lett.* 83, 1602 (1999).

[92] G. E. Morfill, H. M. Thomas, and M. Zuzic, in *Advances in Dusty Plasmas*, edited by P. K. Shukla, D. A. Mendis, and T. Desai (World Scientific, Singapore, 1997), p. 99-142.

[93] R. L. Merlino, A. Barken, C. Thompson, and N. D'Angelo, *Phys. Plasmas* 5, 1607 (1998).

[94] M. Salimullah, *Phys. Lett. A* 215, 296 (1996).

[95] M. K. Islam, M. Salahuddin, A. K. Banerjee, and M. Salimullah, *Phys. Plasmas* 9, 2971 (2002).

[96] M. K. Islam, Y. Nakashima, K. Yatsu, and M. Salimullah, *Phys. Plasmas* 10, 591 (2003).

[97] G. Medicus, *J. Appl. Phys.* 32, 2512 (1961).

[98] L. Scott, Electrical Probes, in *Plasma Diagnostics*, edited by W. Lochte-Holtgreven, (American Vacuum Society Classics, AIP Press, New Work, 1995), P.675.

[99] E. C. Whipple, *Rep. Prog. Phys.* 44, 1197 (1981).

[100] M. R. Jana, A. Sen, and P. K. Kaw, *Phys. Rev. E* 48, 3930 (1993).

[101] N. D'Angelo, *Phys. Lett. A* 292, 195 (2001).

In: Advances in Plasma Physics Research, Volume 7
Editor: Francois Gerard

ISBN: 978-1-61122-983-7
© 2011 Nova Science Publishers, Inc.

Chapter 5

MATERIAL PROBE ANALYSIS FOR PLASMA FACING SURFACE IN THE LARGE HELICAL DEVICE

T. Hino[1], Y. Nobuta[1], N. Ashikawa[2], N. Inoue[2], A. Sagara[2], K. Nishimura[2], Y. Yamauchi[1], Y. Hirohata[1], N. Noda[2], N. Ohyabu[2], A. Komori[2], O. Motojima[2] and LHD Experimental Group[2]

[1] Laboratory for Plasma Physics and Engineering, Hokkaido University, Sapporo, 060-8628 Japan
[2] National Institute for Fusion Science, Toki-shi, Gifu-ken, 509-5202 Japan

ABSTRACT

In the large helical device, LHD, material probe study has been conducted since the first experimental campaign in Mar. 1998. Material probes have been installed at the inner walls along the poloidal direction from the first experimental campaign, and both at the inner wall along the poloidal direction and at the first wall along the toroidal direction from the 4th experimental campaign. After each campaign, the surface morphology, the impurity deposition and the gas retention were examined by using surface analysis techniques in order to clarify the plasma surface interactions and the degree of wall cleaning. In the first experimental campaign, the iron oxide layer at surface was observed to be thick. However, in the 2nd campaign, the entire wall was thoroughly cleaned by glow discharge conditioning and the increase of main discharge shots. From the 3rd campaigns, graphite tiles were installed over the entire divertor strike region, and then the wall condition was significantly changed compared with the case of a stainless steel wall. It was seen that graphite tiles in the divertor were eroded mainly during main discharges, and the stainless steel first wall mainly during glow discharges. The eroded carbon during main discharges was deposited on the entire wall. The fraction of carbon coverage in the first wall was approximately 60%. The deposition thickness of carbon was large at the wall far from the plasma. The reduction of metal impurities in the plasma was observed, which corresponds to the carbonized wall. Since the entire wall was carbonized, the amount of retained discharge gases such as H and He became large. In particular, the helium retention was large at the position close to the anodes used for helium glow discharge cleanings. One characteristic of the LHD wall is a large retention of helium since the wall temperature is limited to below 368 K. From the 5th experimental campaigns, boronization was several times conducted in each campaign to control the gas retention in the wall. The fraction of boronized wall to the entire wall was 20-30%. A large reduction of oxygen impurity level in the plasma was observed after the boronization. This result shows that the oxygen was well trapped in the boron layer even if the coverage of boron was small. The results on the material probes have been referred for the next experimental campaign, and then the plasma confinement was significantly improved. At present, the plasma density limit exceeded 10^{20}m^{-3}, both electron and ion temperatures 10keV and the average beta 4%. These parameters are comparable with those of existing large tokamaks such as JT-60U and JET.

1. INTRODUCTION

Energy balance of fusion plasma is determined by energy confinement of plasma and radiation loss from the plasma. The energy confinement is affected by fuel hydrogen recycling, and the fuel hydrogen retention offers a capability of fuel hydrogen recycling. In a DT fusion reactor, in-vessel tritium inventory is determined by the fuel hydrogen retention. The light mass impurities such as C and O and metal impurities emit into the plasma and then these impurities enhance the radiation loss power. Thus, the fuel hydrogen retention and the degree of impurity emission have to be investigated. Material probe study is very useful to know the impurity deposition, gas retention and erosion of plasma facing wall. In the present fusion devices, only one or several material probes were installed at the inner wall and so that overall behavior of wall surface was not systematically investigated so far.

In the large helical device, LHD, a largest device as a helical confinement device in the world, the material probes were installed at the inner wall along the poloidal direction from the first experimental campaign conducted in Mar. 1998, and at the first wall along the toroidal direction in the 4th campaign and from the 6th experimental campaign. Thus, overall change of the wall surface was investigated and this result was applied for the wall conditioning in the next experimental campaign. After each campaign, the material probes were extracted from the vacuum vessel, and then analyzed by using surface analysis techniques. The impurity deposition on the wall was analyzed by using Auger electron spectroscopy, AES, the surface morphology by scanning electron microscope, SEM, and gas retention by thermal desorption spectroscopy, TDS. The data of the impurity deposition on plasma facing walls and the retention of discharge gas in the wall are useful to understand the impurities in main discharge plasma and the degree of fuel gas recycling. These wall condition data were systematically accumulated as a database and analyzed for the wall characteristics through seven experimental campaigns. In LHD, the wall temperature is limited to below 368 K for the thermal insulation between the vacuum chamber and super-conducting helical coils. Thus, these data are particularly important for the particle control. This paper presents the results obtained in period from the first to 6th campaigns.

2. PROGRESS OF PLASMA PARAMETERS IN LHD

He ECR discharge cleaning was employed in the first campaign for the initial ECH plasma production, and glow discharge cleanings from the 2nd campaign for the production of NBI heated plasmas [1-6]. From the 3rd campaign for ICRF heated plasmas and high-power plasma production, graphite tiles have been installed in the divertor leg region to reduce metal impurities in the plasma [7, 8, 9, 10]. Improved plasma performance was investigated in the 4th campaign, relevant mainly to the magnetic axis position. Boronization was conducted from the 5th campaign to control the gas retention and oxygen impurity. Figure 1 shows the plasma stored energy as a function of shot number from the first to 7th campaigns, representing the progress of LHD plasma performance. The highest values of plasma parameters and magnetic field achieved from the first to 4th campaigns are summarized as follows:

(1) T_e of 1.3 keV, n_e of 1.3×10^{19} m^{-3} and B of 1.5 T in the first campaign,
(2) T_e of 2.3 keV, T_i of 2.0 keV, n_e of 7×10^{19} m^{-3}, averaged beta $<\beta>$ of 1% ($B = 1.5$ T) and B of 2.8 T in the 2nd campaign,
(3) T_e of 4.4 keV, T_i of 3.5 keV, n_e of 1.1×10^{20} m^{-3}, $<\beta>$ of 2.4% (B = 1.3 T), and B of 2.8 T in the 3rd campaign, and
(4) T_e of 4.4 keV, T_i of 3.5 keV, n_e of 1.5×10^{20} m^{-3}, and $<\beta>$ of 3% (B = 0.5 T) and B of 2.8 T in the 4th campaign.

The above parameter sets are not data obtained simultaneously in a single shot. The numbers of main discharge shots and total discharge times are 1888 shots in the first campaign, 5243 shots and 1500 s in the 2nd campaign, 10178 shots and 3200 s in the 3rd campaign, and 8895 shots and 8600 s in the 4th campaign.

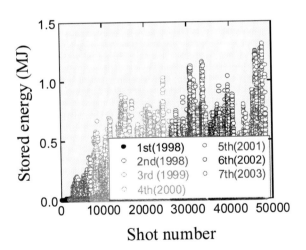

Figure 1 Plasma stored energy with shot number.

From the 4th campaign to 7th campaign, the plasma parameters achieved the values comparable with those of existing large tokamaks such as JT-60U and JET. Namely, the plasma stored energy exceeded 1MJ, average beta 4%, both ion and electron temperatures 10keV, and plasma density 10^{20}m^{-3}.

From the first campaign to the 7th campaign, material probes of SS and graphite were placed on several inner wall positions of the same poloidal cross-section at the toroidal sector

#7 (Figure 2). It is worthwhile, for studying the wall characteristics, to fix the probe positions through the seven campaigns. In the glow discharge cleaning, two anodes were employed. In the 4th, 6th and 7th campaigns, material probes were placed along the toroidal direction as shown in Figure 3, in order to clarify the effect of glow discharge cleaning and boronization, in addition to the probes shown in figure 2. In the 4th campaign, other material probes were also placed at the port, and the probe samples exposed to only main discharges and to only glow discharges were prepared, in order clarify the plasma surface interactions only during main discharge and glow discharge. After each campaign, impurity deposition, change of surface morphology, and retention properties of discharge gas and impurity gas were examined in order to clarify the change of wall surfaces and degree of wall cleaning.

Toroidal sector #7

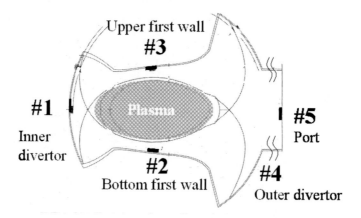

Figure 2 Position of material probes at the inner wall of #7 toroidal sector.

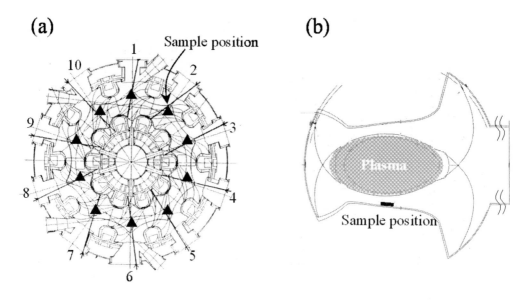

Figure 3 Toroidal positions of material probes (a) and poloidal position of material probe (b).

3. PLASMA FACING SURFACE FOR PERIOD FROM FIRST CAMPAIGN TO THIRD CAMPAIGN

After the first campaign, the change of surface morphology and surface atomic composition were examined by using SEM and AES, respectively. On the entire wall surface the deposition of many sub-micron particles was observed. Figure 4 shows the photographs of graphite samples taken by SEM for cases before and after the exposure. After the exposure (Figure 4 (b) and (c)), the shape of graphite matrix or binder was rounded by the impurity deposition. The atomic composition of the surface was analyzed by AES. The deposition was identified as Fe-O particles [11]. The depth profile of surface atomic composition was also taken and the content of impurities such as O and C were investigated. The concentration of oxygen was observed to be large, and the oxygen concentration and the deposition thickness were 60 at.% and 200 nm, respectively. The retained amounts of discharge gas such as H_2 and He and the retained amounts of impurity gases such as H_2O, CO and CO_2 were measured by using a technique of TDS [12]. In the TDS apparatus, the sample was heated with a constant heating rate. The desorption rate was measured during the heating by using a quadruple mass spectrometer, QMS. The retained amount was obtained by integrating the desorption rate with respect to heating time. The amounts of these retained gases were large at the walls far from the plasma, the outer divertor region #4 and the port #5. The temperature rise during the discharge was very small in the entire wall. In the first campaign, He ECR discharge cleaning was conducted as the wall conditioning method, and the number of main discharge shots was only 1888. Thus these results suggest that the walls were not cleaned yet by such discharges. However, the gas retention and the impurity concentration at the surface were relatively small at the wall near the plasma such as upper and bottom first walls, #2 and #3. Then, the ECR discharge cleaning and/or the main discharge were effective for the wall near the plasma but not for the wall far from the plasma.

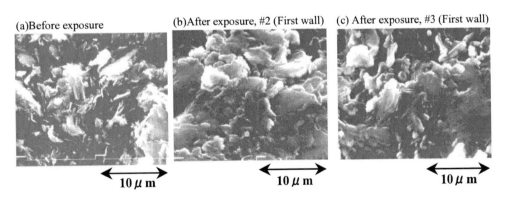

Figure 4 Surface morphologies of graphite sample before the exposure (a) and after the exposure (b) and (c).

In the 2nd campaign, glow discharge cleaning was conducted instead of ECR discharge, for the walls far from the plasma to be cleaned. The number of main discharge shot increased to approximately 5000 shots. After the 2nd campaign, the deposition of Fe-O sub-micron particles disappeared at the wall except in the inner divertor leg region. In addition, the oxygen concentration at surface became low, 40 at.% and the deposition thickness became thin, 20 nm. The reduction of oxygen impurity at the surface was significantly large. The total gas retention also decreased by 30%, in particular, the decrease was largest at the wall far from the plasma. In the 2nd campaign, the retention of discharge gases, H_2 and He, was clearly observed in every material probe. The retention of He, the gas species used for main discharges and glow discharges, is not small, compared with the hydrogen retention. In other large fusion devices, such a large retention of He has not been observed so far. The large retention of He is one of characteristics of LHD walls. It is known that the helium is significantly retained in the wall when the wall temperature is low [13]. In the LHD, the wall temperature is as high as 368 K. Thus the retention of He might have been large owing to the low wall temperature. Compared to the first campaign, the impurity concentration at the wall surface and the retained amount of impurity gases were significantly reduced. These results suggest that both the glow discharge cleaning and the increase of main discharge shots were effective for the wall conditioning.

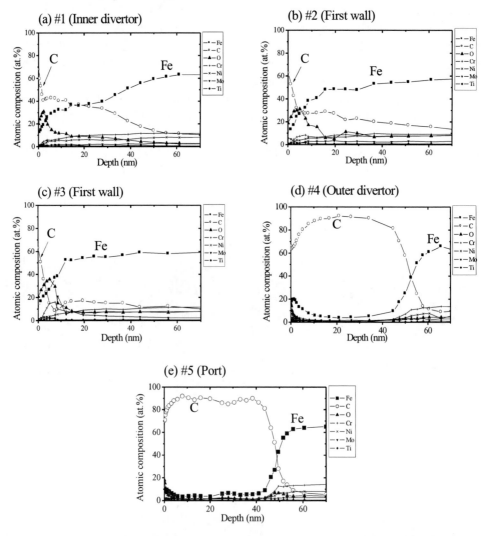

Figure 5 Depth profiles of atomic composition near the inner divertor (a), at the first wall (b) and (c), near the outer divertor (d) and the port (e).

From the 3rd campaign, the wall surface was substantially changed by the installation of graphite tiles at the divertor leg region. The purpose of the installation of graphite tiles is the protection of the stainless steel wall in the divertor region from the heat flux of the high power discharge and the reduction of metal impurity levels in the plasma. Figure 5 shows the depth profiles of atomic composition in the samples at near the inner divertor (a), the first walls (b) and (c), near the outer divertor leg regions (d) and at the port (e). The entire wall was covered by carbon. In the first walls (b) and (c), the carbon concentration at the top surface was approximately 60 at.%. The carbon concentration was high, 90 at.% at the walls far from the plasma, (d) and (e). It was observed that the impurity concentration of Fe in the plasma was reduced to approximately a half of the 2nd campaign in the early stage of the 3rd campaign [7, 8]. A large reduction of Fe impurities is due to the carbonized wall. The concentrations of oxygen and carbon increased in the early stage of the 3rd campaign, but these levels decreased to the levels of the 2nd campaign with the increase of the shot number. Although the graphite contains a large amount of impurities such as H_2, H_2O, CO_2 and CO [14], these impurities might have been mostly desorbed in the later stage of the 3rd campaign. Figure 6 shows the amounts of gas retained in samples after the 3rd campaign. The amount of retained gas doubled compared with the case of the 2nd campaign. The increase of the gas retention is due to the deposition of carbon, since the carbon is porous and therefore the gas absorption area, effective surface area, is quite large compared to the geometric area [14]. In particular, the retained amount at the wall far from the main plasma increased greatly owing to the thick deposition of carbon. Figure 7 shows the amount of retained helium for the positions shown in Figure 2. The helium gas was employed for half of the main discharge shots and helium glow discharge cleanings. The helium retention was observed at the entire wall also in the 3rd campaign. The helium retention was large at the wall close to the plasma. It is presumed that the helium retention took place owing to implantation of charge exchanged helium during the main helium discharge and helium ions during the helium glow discharge. The energies of charge exchanged helium and helium ion in the glow discharge are several keV and approximately 300 eV, respectively.

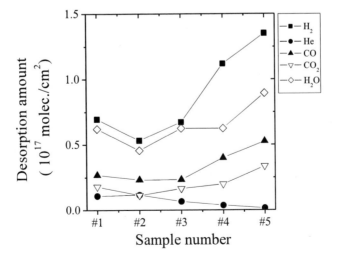

Figure 6 Amounts of gas retained in the samples after the 3rd campaign.

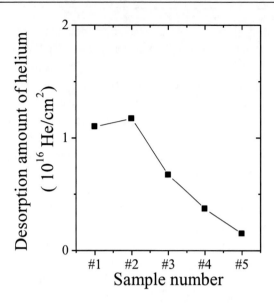

Figure 7 Amounts of helium retained in the samples installed along the poloidal direction at toroidal sector #7 after the 3rd campaign.

4. PLASMA FACING SURFACE FOR PERIOD OF FOURTH CAMPAIGN

In the 4th campaign, material probes were exposed to only the main discharge or only to the glow discharge to understand the helium retention. In addition, the probes were installed along the toroidal direction to examine the toroidal dependence of the helium retention.

The surface of the material probes placed along the poloidal direction at toroidal sector #7 after the 4th campaign was also carbonized, similar to that after the 3rd campaign. It is noted that the toroidal sector #7 shown in Figure 2 is far from the anode used for the helium glow discharge. The material probes installed along the toroidal direction are shown in Figure 3. The poloidal position of the sample was at the bottom first wall and the number of the samples was 10. In Figure 8, the positions of the anodes for glow discharge, ECH, NBI and ICRH are shown. The amounts of retained helium and retained impurity gas in these samples are shown in Figure 9. The sample close to the anodes retained a large amount of helium. The retention of hydrogen showed the same tendency. Since the current density of the glow discharge in the vicinity of the anode is higher than the toroidal region far from the anode, the large retention of helium is due to ion implantation during helium glow discharge. The deposition of impurity gas was also large in the sample close to the anodes. The deposition of Fe showed a similar tendency. It is presumed that the impurities emitted during glow discharge were ionized in the dense plasma near the anode and deposited to the wall. The relation of the impurity deposition and gas retention on the heating positions of ECH, NBI and ICRH was not clearly observed.

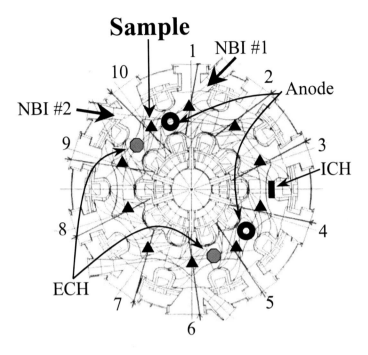

Figure 8 Positions of anodes and heating ports.

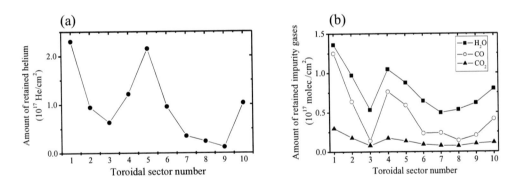

Figure 9 Amounts of helium (a) and impurity gases (b) retained in the samples placed along the toroidal direction.

The samples exposed to only main discharge shots or glow discharge cleaning were prepared using a rotating shutter, in addition to the samples along the poloidal and toroidal directions. Figure 10 shows the depth profiles of atomic composition for a stainless steel (SS) sample exposed to only glow discharges (a) and main discharges (b). Figure 11 shows depth profiles of atomic composition for a graphite sample exposed to only helium glow discharges (a) and main discharges (b). In the SS sample exposed to only main discharges, carbon deposition was dominant, owing to erosion of the divertor tiles. This behavior can be understood as follows. During the main discharge, the carbon eroded at the divertor is ionized and transported to the other regions along magnetic field lines, and deposited on the wall at the end of the discharge. On the other hand, in the graphite sample exposed to only the glow

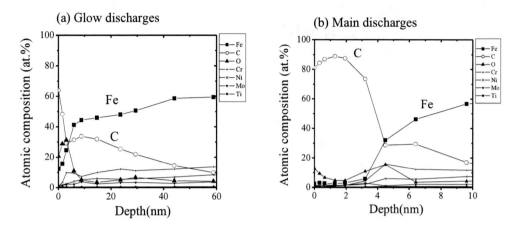

Figure 10 Depth profiles of atomic composition for the SS samples exposed to only glow discharges (a) and to only main discharges (b).

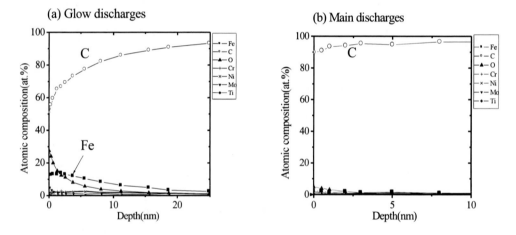

Figure 11 Depth profiles of atomic composition for the graphite samples exposed to only glow discharges (a) and to only main discharges (b).

discharges, dominant Fe deposition was observed. The atomic concentration of Fe at the first wall was approximately 40 at.% and the current density of the glow plasma between the first wall and anode is higher than that between the graphite tiles and anode. Then the sputtering of Fe becomes relatively dominant in the case of glow discharge. The dominant deposition of Fe can be explained by this reason. For these samples, the TDS analysis was conducted and the desorption spectrum of He was obtained. In this analysis, the sample extracted from the vacuum chamber was heated from room temperature to 1273 K in the TDS chamber. During the heating, the emitted gas was quantitatively measured by using quadruple mass spectrometer, QMS. Figure 12 shows the desorption spectra of helium retained in the SS samples exposed to only glow discharges and only main discharges (a) and in the graphite samples exposed to only glow discharges and only main discharges (b). The vertical and horizontal axes show the desorption rate and heating temperature, respectively. In the SS sample exposed to only glow discharges, the retained helium desorbed in the temperature regime lower than approximately 800 K. In the SS sample exposed to only main discharges, the retained helium desorbed in the regime higher than approximately 800 K. It is known that the desorption temperature increases with the increase of the helium ion energy [15]. The charge exchanged helium during the main discharge has an energy higher than that of helium ions in the glow discharge. The energy of helium ions in the glow discharge is as high as approximately 300 eV. Figure 12 shows the results for the samples placed at the wall far from the anodes. In this case, the helium retention owing to the charge exchanged helium is comparable to that owing to helium ions of glow discharge. The contribution of the glow discharge is very large for the samples close to the anodes as shown in Figure 8. Thus, the major contribution to the helium retention is the He glow discharges. In the case of a hydrogen main discharge, the contamination of helium into the hydrogen plasma was observed. Thus, the hydrogen glow discharge has been conducted before the hydrogen main discharge to reduce the contamination of helium. The desorption spectra shown in Figure 12 indicates that the helium retention can be significantly reduced if the wall is baked with a temperature of approximately 800K. Hence, the surface baking is desirable for reduction of helium recycling.

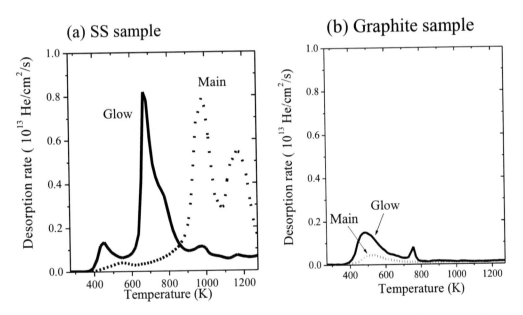

Figure 12 Desorption spectra of helium for the SS samples exposed to only glow discharges and to only main discharges (a), and for the graphite samples exposed to only glow discharges and to only main discharges (b).

5. PLASMA FACING SURFACE FOR PERIODS OF FIFTH AND SIXTH CAMPAIGNS

In the 5th and 6th campaigns, boronization was 3 times conducted in each campaign in order to control oxygen impurities and retention of discharge gas. The glow discharge using a mixture gas consisting diborane and helium was employed for the boronization [16]. Figure 13 shows positions of anodes and gas inlets. In the 6th campaign, the gas flow at the inlets between toroidal sectors 1 and 2 was largest and that at the toroidal sectors between 7 and 8 smallest. Figure 14 shows the thickness of boron film and trapped amount of oxygen in the boronized wall versus toroidal sector number. In the sample at the toroidal sectors 1 and 10, close to the anode and gas inlet with a large flow, the boron thickness were large, 400nm and 125nm, respectively. The boron concentration at these positions was high, 80%, as shown in Figure 15. At the position of toroidal sector 4, the boron thickness was 50nm and the concentration was approximately 50%. In the other positions, the thickness was only several nm and the concentration several percents. The coverage of the boron in the entire wall was approximately 25-30%. From Figure 14, it is seen that the oxygen amount trapped by the boron layer roughly corresponds to the boron thickness. Namely, the oxygen was well trapped in the thick boron layer.

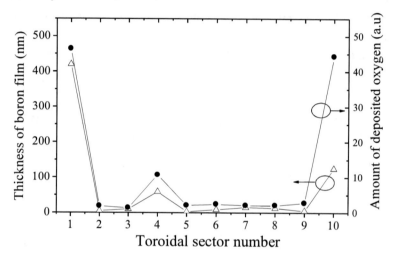

Figure 13 Positions of anodes and gas inlets for boronization.

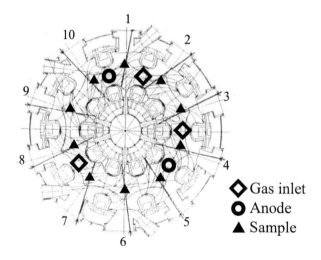

Figure 14 Toroidal dependence of boron thickness and trapped amount of oxygen.

In the 6th campaign, the oxygen impurity and carbon impurity levels in the plasma became 1% and 40% of the levels before the boronization [17]. The radiation loss power was 50% reduced compared to that before the boronization. The present boronization was very useful for the impurity control although the wall coverage was only 20-30%. The retention of helium and hydrogen in the boronized wall was analyzed by using TDS. The retained amount of helium was similar to that of stainless steel wall but the retained amount of hydrogen became smaller than that of stainless steel. Hence, the boronized wall may contribute for the reduction of fuel hydrogen recycling.

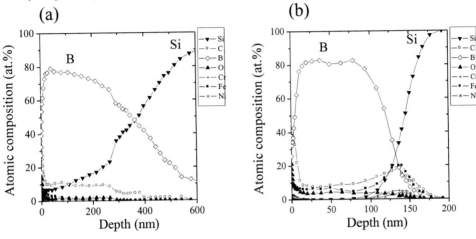

Figure 15 Depth profiles of atomic composition for the samples at toroidal sector 1 and 10.

6. SUMMARY AND CONCLUSION

In order to understand systematically the change of plasma facing walls in LHD, material probes were installed and analyzed for the period from the first experimental campaign to the 6th experimental campaign. The material probes were installed along the poloidal and toroidal direction to investigate the positional dependence of plasma surface interactions. In LHD, discharge cleanings such as glow discharge and ECR discharges have been employed, and the gas species were hydrogen and helium. As the main discharge, both the hydrogen and helium discharges were conducted so far. The fraction of helium main discharges was roughly the same as the hydrogen main discharges. After the each campaign, the change of the surface atomic composition and surface morphology, and the retention of discharge and impurity gases were analyzed using surface analysis techniques. In particular, the impurity deposition on the wall and the gas retention of the wall surface were examined.

After the first campaign, many sub-micron particles were deposited on the entire wall and these were identified as Fe-O particles. In the first and 2nd campaigns, the material of all walls was stainless steel, SS 316L. The oxygen impurity was dominant at the surface, and the concentration and oxide thickness were 60 at.% and 200 nm, respectively. In particular, the amount of retained impurities was large in the walls relatively far from the plasma. In the first campaign, ECR discharge cleanings and approximately 2000 main discharges were conducted. The result indicates that the wall was not sufficiently cleaned by such discharges.

In the 2nd campaign, glow discharge cleaning was employed instead of ECR discharge for the walls far from the plasma to be cleaned. The number of main discharge shots increased to approximately 5000. After the campaign, the deposition of Fe-O particle mostly disappeared and the oxygen concentration and oxide thickness were decreased to 40 at % and 20 nm, respectively. The amount of retained impurity and discharge gases decreased by 30 %. In particular, the gas retention of the wall far from the plasma significantly decreased. This reduction is mainly due to the glow discharge cleaning. The retention of hydrogen and helium was clearly observed from the 2nd campaign. The helium retention is not negligible, compared with the hydrogen retention. Since the helium retention becomes large when the wall temperature is low, such a large helium retention is due to the low wall temperature of LHD.

From the 3rd campaign, the graphite tiles were installed in the divertor leg regions. The wall surface was significantly changed by the deposition of carbon. The carbon concentration became 60 at.% at the first wall and 90 at.% at the outer divertor and the port. The Fe impurity level in the plasma was reduced to approximately half of the 2nd campaign. This reduction is consistent with the carbonized wall. On the other hand, the gas retention was doubled since the discharge gas and impurity gas is well retained by the carbon. At the wall far from the plasma, the gas retention was larger than that of the wall close to the plasma since the carbon layer was thicker. The helium retention was more clearly observed in the entire wall.

In the 4th campaign, the material probes were exposed to only main discharges and only glow discharges. In the probe exposed to only main discharges, a large amount of carbon was deposited. In the probe exposed to only glow discharges, the deposition of iron was relatively large. These suggest that the plasma surface interactions during the main discharge and the glow discharge took place mainly at the divertor and the first wall close to the plasma, respectively. The toroidal dependence of the impurity deposition, helium and hydrogen retention was investigated by analyzing the probes placed along the toroidal direction. The impurity deposition and retention of hydrogen and helium were large at the wall in the vicinity of the anodes. Thus the glow discharge plasma significantly changed the condition of the wall surface. Helium was retained from both helium main discharges and helium glow discharges. Of the helium retention, the contribution owing to the helium glow discharge was superior in the present LHD condition.

In the 5th and 6th campaigns, boronization was several times conducted by using a glow discharge with a mixture gas of diborane and helium in each campaign. The thickness of boron film was large at the wall close to the anode and gas inlet. The amount of trapped oxygen increased as the boron thickness. In the plasma, the oxygen impurity level was two orders of magnitude reduced after the boronization. The wall data well corresponds to the reduction of oxygen impurity level. The hydrogen retention decreased in the boronized wall compared with the case of stainless steel, and thus the fuel hydrogen recycling might have been suppressed by the boronization.

The wall behavior in LHD was characterized, corresponding to the progress of plasma performance with the increase of heating power and the installation of graphite tiles for the divertor. The graphite tiles were installed over the entire region of the divertor trace, and then the wall condition was significantly changed compared with the case of the previous SS wall. The entire wall was well carbonized, after which a large reduction of the Fe impurity level in the plasma was observed. However, the gas retention was increased by the deposition of carbon. The retention of discharge gases such as helium is not negligible and is one of the characteristics of the LHD wall. The large retention of helium is due to the low wall temperature. One method of reducing the gas retention is baking such as wall surface heating between the main discharge shots. Since the desorption temperatures of hydrogen and helium are 250°C and 300°C, respectively [18], the required baking temperature is approximately 300°C. The other method is to apply the glow discharge with heavy inert gas species such as argon and neon. Since the mass impact and energy transfer of these ions is large, the retained helium may be desorbed and/or removed by the sputtering. For reduction of oxygen impurities, boronization was observed to be effective in the 5th and 6th campaigns. Both the wall conditionings by baking and glow discharge with heavy inert gas species and the boronization will lead to further improvement of the LHD plasma.

REFERENCES

[1]   Fujiwara, M. et al. 2001 *Nucl. Fusion* 41 1355

[2]   Motojima, O. et al. 2003 *Nucl. Fusion* 43 1674

[3]   Sagara, A. et al. 1999 *J. Plasma and Fusion Res.* 75 263

[4]   Inoue, N. et al. 2000 *J. Plasma and Fusion Res. Series* Vol. 3 324

[5]   Hino, T. et al. 2001 *J. Nucl. Mater.* 290-293 1176

[6]   Masuzaki, S. et al. 2001 *J. Nucl. Mater.* 290-293 12

[7]   Peterson, B.J. et al. 2001 *J. Nucl. Mater.* 290-293 930

[8]   Morita, S. et al. 2001 *Phys. Scripta* T91 48

[9]   Hino, T. et al. 2003 *J. Nucl. Mater.* 313-316 167

[10] Sagara, A. et al. 2003 *J. Nucl. Mater.* 313-316 1

[11] Hino, T. et al. 2001 *J. Nucl. Mater.* 290-293 1176

[12] Hino, T. et al. 2000 *Fusion Eng. and Design* 49-50 213

[13] Hino, T. et al. 1999 *J. Nucl. Mater.* 266-269 538

[14] Yamashina, T. and Hino, T. 1990 *J. Nucl. Sci. Technol.* 27 589

[15] Nobuta, Y., Hino, T. et al. 2003 *J. Vac. Soc. Jpn.* 46 628

[16] Nobuta, Y., Hino, T. et al. "Impurity Deposition and Retention of Discharge Gas on Plasma Facing Wall in LHD", 2003 Presented in 11th International Conference on Fusion Reactor Material, Dec. 8-12, 2003, Kyoto

[17] Nishimura, K. et al, 2003 *J. Plasma and Nuclear Fusion Res.* 12 1216

[18] Hino, T. et al, 2004 Submitted to *Fusion Eng. Design*

In: Advances in Plasma Physics Research, Volume 7
Editor: Francois Gerard

ISBN: 978-1-61122-983-7
© 2011 Nova Science Publishers, Inc.

Chapter 6

NONLINEAR DYNAMICS OF SELF-GUIDING ELECTROMAGNETIC BEAMS IN RELATIVISTIC ELECTRON-POSITRON PLASMAS

M. Ohhashi[1], T. Tatsuno[1,2], V.I. Berezhiani[3] and S.V. Mikeladze[3]
[1]Graduate School of Frontier Sciences,
University of Tokyo, Hongo 7-3-1, Tokyo 113-0033, Japan
[2]Institute for Research in Electronics and Applied Physics,
University of Maryland, College Park, Maryland 20742-3511, USA
[3]Institute of Physics, The Georgian Academy of Sciences,
6 Tamarashvili str., Tbilisi 380077, Georgia

Abstract

Nonlinear interaction of an intense electromagnetic (EM) beam with relativistically hot electron-positron plasma is investigated by invoking the variational principle and numerical simulation, resting on the model of generalized nonlinear Schrödinger equation with saturating nonlinearity. The present analysis shows the dynamical properties including the possibilities of trapping and wave-breaking of EM beams. These properties of EM beams may give a significant clue for the gamma-ray burst.

1 Introduction

The problem of electromagnetic (EM) wave propagation and related phenomena in relativistic plasma have attracted considerable attention in the recent past. Relativistic electron-positron (e-p) dominated plasmas are created in a variety of astrophysical situations. Electron-positron pair production cascades are believed to occur in pulsar magnetospheres [1]. The e-p plasmas are also likely to be found in the bipolar outflows (Jets) in Active Galactic Nuclei (AGN) [2], and at the center of our own Galaxy [3]. In AGNs, the observations of superluminal motions are commonly attributed to the expansion of relativistic e-p beams in a pervading subrelativistic medium. This model implies copious pair production via γ-γ interactions creating an e-p atmosphere around the source. The actual production of e-p pairs due to photon-photon interactions occurs in the coronas of AGN accretion disks, which upscatter the soft photons emitted by the accretion disks by inverse Compton scattering. The presence of e-p plasma is also argued in the MeV epoch of the early Universe [4]. On the other hand the contemporary progress in the development of super-strong laser sources with intensities $I = 10^{21-23}$ W/cm^2 has also made it possible to create relativistic e-p plasmas in the laboratory by a variety of experimental techniques [5]. Elucidation of the electromagnetic wave dynamics in a relativistic e-p plasmas will, perhaps, be an essential determinant of the radiation properties of astrophysical objects as well as of the medium exposed to the field of super-strong laser radiation.

Wave self-modulation and soliton-formation is, perhaps, one of the most interesting and significant features of the overall plasma dynamics. The existence of stable localized envelope solitons of EM radiation has been suggested as a potential mechanism for the production of micro-pulses in AGN and pulsars [6, 7, 8]. In the early Universe localized solitons are strong candidates to explain the observed inhomogeneities of the visible Universe [9, 10], and vortex soliton can be considered to play a significant role for the formation of the observed structure of the Universe [11]. The gamma-ray bursts (GRBs) and their afterglows are likely to be the result of energy dissipation from a relativistically expanding outflow [12]. A Poynting-flux driven outflow from a magnetized rotator is a promising paradigm for GRB engines and there have been various implementations of this concept [13, 14, 15, 16, 17]. Recently Lyutikov and Blackman suggested that gamma-rays are emitted at the point where MHD breaks down due to the overturn instability of large amplitude electromagnetic wave [18].

In the present paper we explore the mechanism for the localization of multi-dimensional intense EM radiation in pure e-p plasmas. Assuming the plasma to be transparent to the beam, and applying a fully relativistic hydrodynamical model, we demonstrate the possibility of beam self-trapping leading to the formation of stable 2D solitonic structures. The high-frequency pressure force of the EM field (tending to completely expel the pairs radially from the region of localization) is overwhelmed by the thermal pressure force which opposes the radial expansion of the plasma creating conditions for the formation of the stationary self-guiding regime of beam propagation.

2   Basic Equations

In this section we apply our general formulation to the problem of self-trapping of EM beams in pure e-p plasmas with relativistic temperatures. We assume that the equilibrium state of the plasma is characterized by an overall charge neutrality $n_\infty^- = n_\infty^+ \equiv n_\infty$, where n_∞^- and n_∞^+ are the unperturbed number densities of the electrons and positrons in the far region of the EM beam localization. In most mechanisms for creating e-p plasmas, the pairs appear simultaneously and due to the symmetry of the problem it is natural to assume that $T_\infty^- = T_\infty^+ \equiv T_\infty$, where T_∞^- and T_∞^+ are the respective equilibrium temperatures.

We shall assume that for the radiation field of interest, the plasma is underdense and transparent, i.e., $\epsilon = \omega_e/\omega \ll 1$, where ω is the mean frequency of EM radiation and $\omega_e = (4\pi e^2 n_\infty/m_{0e})^{1/2}$ is the plasma frequency. Following Ref. [19] we introduce a temperature dependent momentum $\mathbf{\Pi}^\pm = G^\pm \mathbf{p}^\pm$ and relativistic factor $\Gamma^\pm = G^\pm \gamma^\pm$, where $G = K_3(z)/K_2(z)$ is the temperature dependent factor with K_n the n-th order modified Bessel function of the second kind ($z = m_{0e}c^2/T$), $\mathbf{p}^\pm$ and $\gamma^\pm = \sqrt{1 + (\mathbf{p}^\pm/m_e c)^2}$ are

N

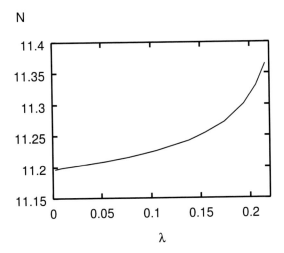

Figure 1: The beam power N versus λ ($T_\infty = 1$).

respectively momentum and relativistic factor of e-p particles. Introducing the dimensionless quantities $\tilde{t} = \omega t$, $\tilde{\mathbf{r}} = (\omega/c)\mathbf{r}$, $\tilde{T}^\pm = T^\pm/m_0 e c^2$, $\tilde{\mathbf{A}} = e\mathbf{A}/(m_0 e c^2)$, $\tilde{\phi} = e\phi/m_0 e c^2$, $\tilde{\mathbf{\Pi}}^\pm = \mathbf{\Pi}^\pm/(m_0 e c)$, and $\tilde{n}^\pm = n^\pm/n_\infty$, we arrive at the dimensionless equations [19],

$$\frac{\partial \mathbf{\Pi}^\pm}{\partial t} + \nabla \Gamma^\pm = \mp \frac{\partial \mathbf{A}}{\partial t} \mp \nabla \phi, \tag{1}$$

$$\frac{n^\pm}{\Gamma^\pm f(T^\pm)} = \text{const}, \tag{2}$$

$$\frac{\partial n^\pm}{\partial t} + \nabla \cdot \mathbf{J}^\pm = 0, \tag{3}$$

$$\Delta \phi = \epsilon^2 (n^- - n^+), \tag{4}$$

$$\frac{\partial^2 \mathbf{A}}{\partial t^2} - \Delta \mathbf{A} + \frac{\partial}{\partial t} \nabla \phi - \epsilon^2 (\mathbf{J}^+ - \mathbf{J}^-) = 0, \tag{5}$$

where

$$f(T^\pm) = \frac{K_2(1/T^\pm)T^\pm}{G^\pm(T^\pm)} \exp[G^\pm(T^\pm)/T^\pm], \tag{6}$$

with $\mathbf{J}^\pm = n^\pm \mathbf{\Pi}^\pm/\Gamma^\pm$ and $\Gamma^\pm = \sqrt{(G^\pm)^2 + (\mathbf{\Pi}^\pm)^2}$. Here $\mathbf{A}$ and ϕ are the vector and scalar potentials ($\nabla \cdot \mathbf{A} = 0$). The tilde is suppressed for convenience.

Of various techniques that could be invoked to investigate Eqs. (1)-(6) to study the self-trapping of high-frequency EM radiation propagating along the z-axis, we choose the method presented in the excellent paper by Sun et al. [20]. The method is based on the multiple scale expansion of the equations in the small parameter ϵ. Assuming that all variations are slow compared to the variation in $\xi = z - at$, we expand all quantities $Q = (\mathbf{A}, \phi, \mathbf{\Pi}^\pm, n^\pm, \ldots)$ as

$$Q = Q_0(\xi, x_1, y_1, z_2) + \epsilon Q_1(\xi, x_1, y_1, z_2), \tag{7}$$

where $(x_1, y_1, z_2) = (\epsilon x, \epsilon y, \epsilon^2 z)$ denote the directions of slow change, and $a_1 = (a^2 - 1)/\epsilon^2 \sim 1$. We further assume that the high-frequency EM field is circularly polarized,

$$\mathbf{A}_{0\perp} = \frac{1}{2}(\hat{\mathbf{x}} + i\hat{\mathbf{y}})A\exp(i\xi/a) + \text{c.c.,} \tag{8}$$

where A is the slowly varying envelope of the EM beam, $\hat{\mathbf{x}}$ and $\hat{\mathbf{y}}$ denote unit vectors, and c.c. is the complex conjugate.

We now give a short summary of the steps in the standard multiple-scale methodology (Ref. [20]). In the lowest order in ϵ, the transverse (to the direction of EM wave propagation z) component of Eq. (1) reduces to

$$\mathbf{\Pi}_{0\perp}^{\pm} = \mp\mathbf{A}_{0\perp}. \tag{9}$$

To the next order (in ϵ), the transverse component of Eq. (1) reads:

$$-a\frac{\partial\mathbf{\Pi}_1^{\mp}}{\partial\xi} + \nabla_\perp\Gamma_0^{\pm} = \pm a\frac{\partial\mathbf{A}_1}{\partial\xi} \mp \nabla_\perp\phi_0. \tag{10}$$

Averaging Eq. (10) over the fast variable ξ we obtain $\nabla_\perp\Gamma_0^{\pm} = \mp\nabla_\perp\phi_0$. Using these relations, we obtain $\nabla_\perp\phi_0 = 0$ and

$$\Gamma_0 \equiv \Gamma_0^{\pm} = \text{const.} \tag{11}$$

Note that from the lowest order of Eqs. (2) and (4), we conclude that $n_0^+ = n_0^- \equiv n_0$ and $T_0^+ = T_0^- \equiv T_0$. The relation between EM field and temperature can be found by invoking Eq. (11). Using Eqs. (8) and (9) and determining integration constant such that $A \to 0$ and $T_0 \to T_\infty$, we obtain

$$G^2(T_0) = G^2(T_\infty) - |A|^2. \tag{12}$$

The equation for slowly varying envelope A of EM beam can be obtained from Eq. (5). To the lowest order we find

$$a_1\frac{\partial^2\mathbf{A}_{0\perp}}{\partial\xi^2} - \nabla_\perp^2\mathbf{A}_{0\perp} - 2\frac{\partial^2\mathbf{A}_{0\perp}}{\partial\xi\partial z_2} + 2\frac{n_0(T_0)}{G(T_\infty)}\mathbf{A}_{0\perp} = 0. \tag{13}$$

In deriving this equation, we have used the relation

$$\Gamma_0 = \sqrt{G^2(T_0) + |A|^2} = G(T_\infty). \tag{14}$$

While from Eq. (2), we have

$$n_0(T_0) = \frac{f(T_0)}{f(T_\infty)}. \tag{15}$$

Substituting Eq. (8) into Eq. (13) we find:

$$2i\frac{\partial A}{\partial z} + \nabla_\perp^2 A + \frac{2}{G(T_\infty)}[1 - n(T_0)]A = 0, \tag{16}$$

V(a)

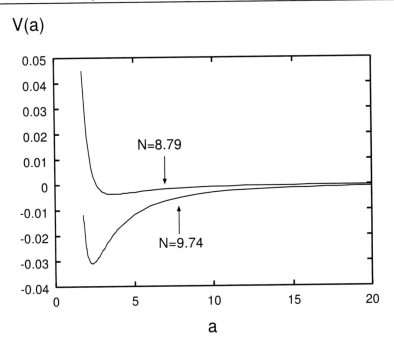

Figure 2: Potential profile of $V(a)$ versus a in the case of $N = 8.79$ and $N = 9.74$.

where subscripts for variables (x_1, y_1, z_2) are dropped for simplicity. We also assumed without loss of generality that $(a^2 - 1)/\epsilon^2 a^2 = 2/G(T_\infty)$, which in dimensional units (provided that $a = \omega/kc$) coincides with the linear dispersion relation of the EM wave in an e-p plasma:

$$\omega^2 = k^2 c^2 + \frac{2\omega_e^2}{G(T_\infty)}. \tag{17}$$

Thus, the dynamics of EM beams in hot relativistic e-p plasma has become accessible within the context of a generalized nonlinear Schrödinger equation (16).

We first seek the localized 2D soliton solutions of Eq. (16), and analyze the stability of such solutions. Making the self-evident re-normalization of variables $z \to zG(T_\infty)$, $r_\perp \to r_\perp \sqrt{G(T_\infty)/2}$, Eq. (16) can be written as:

$$i\frac{\partial A}{\partial z} + \nabla_\perp^2 A + \Psi A = 0, \tag{18}$$

where $\Psi = 1 - n_0(T_0)$ represents the generalized nonlinearity. The companion equation (15) can be viewed as a transcendental algebraic relation between T_0 and $|A|^2$, i.e. Ψ is an implicit function of $|A|^2$ [$\Psi = \Psi(|A|^2)$]. We note that Eq. (18) can be written in the Hamiltonian form $i\partial_z A = \delta H/\delta A^*$, where the Hamiltonian is

$$H = \int d\mathbf{r}_\perp [|\nabla_\perp A|^2 - F(|A|^2)], \tag{19}$$

and $F(t) = \int_0^t \Psi(t')dt'$. This implies that Eq. (18) conserves the Hamiltonian H in addition to the power (photon number) $N = \int d\mathbf{r}_\perp |A|^2$.

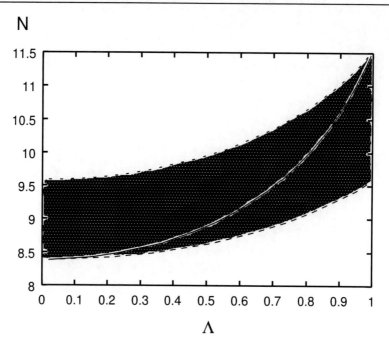

Figure 3: The parameter regime where solitary wave can be trapped for $T_\infty \ll 1$.

To find a stationary, nondiffracting axisymmetric solution, we use the representation of vector potential $A/A_c = U(r) \exp(i\lambda z)$, where $A_c = \left[G(T_\infty)^2 - 1 \right]^{1/2}$, $r = (x^2 + y^2)^{1/2}$ and λ is the nonlinear wave-vector shift. The radially dependent envelope $U(r)$ obeys a nonlinear ordinary differential equation:

$$\frac{d^2U}{dr^2} + \frac{1}{r}\frac{dU}{dr} - \lambda U + \Psi(U^2)U = 0. \tag{20}$$

The profiles of the field $U(r)$, the plasma density $n_0(r)$, and the temperature $T_0(r)$ of the stationary solution can be found in Fig. 3 of Ref. [19] for $\lambda = 0.1$; The plasma temperature and density is reduced in the region of field localization. Similar plots could be obtained for all allowed values of λ. When $\lambda \to \lambda_c$, where λ_c is the upper bound of the propagation constant, the plasma cavitation takes place, i.e. the plasma density and temperature tends down to zero at $r = 0$. Appearance of zero temperature is not surprising since the corresponding region is the "plasma vacuum"; all particles are gone away.

The stability of obtained soliton solutions can be investigated using the stability criterion of Vakhitov and Kolokolov [21]. According to this criterion the soliton is stable against small arbitrary perturbations if

$$\frac{dN}{d\lambda} > 0, \tag{21}$$

where N is the photon number or more precisely the power of the trapped mode. In our case, the dependence of N on λ is shown in Fig. 1. One can see that $dN/d\lambda > 0$ everywhere and consequently the corresponding solution is stable for $0 < \lambda < \lambda_c$.

3 Nonlinear Dynamics Based on Variational Approach

The complex dynamics of a beam governed by Eq. (18) can be analyzed by the variational approach [22]. This approach determines the relations between the characteristic parameters of the localized solution approximated by a trial function. The variational method gives qualitatively good results, provided the beam does not undergo structural changes during its evolution. The first standard step is to construct the Lagrangian

$$L = \frac{1}{2}\left(A^* \frac{\partial A}{\partial z} - \text{c.c.}\right) - \mathcal{H}, \tag{22}$$

where $\mathcal{H}$ is the Hamiltonian density $[H = \int d\mathbf{r}_\perp \mathcal{H}$, see Eq. (19)]. In the optimization procedure, the first variation of the variational function must vanish on a suitably chosen trial function. As a trial function, we will use the Gaussian-shaped beam,

$$A = \Lambda(z) \exp\left[-\frac{r^2}{2a(z)^2} + ir^2 b(z) + i\phi(z)\right], \tag{23}$$

with the amplitude Λ, the beam radius a, the wave front curvature b and the phase ϕ as the unknown functions of the propagation coordinate z respectively, which will be furthermore used in order to make the variational functional an extremum. Substituting expression (23) into Eq. (22) and demanding that the variation of the spatially averaged Lagrangian with respect to each of these parameters is zero, we obtain the corresponding set of Euler-Lagrange equations,

$$\frac{d}{dz}\left(\Lambda^2 a^2\right) = 0, \tag{24}$$

$$\frac{d^2 a}{dz^2} = \frac{4}{a^3} - \frac{2}{a}\left[K'(\Lambda^2) - \frac{K(\Lambda^2)}{\Lambda^2}\right], \tag{25}$$

$$b = \frac{1}{4a}\frac{da}{dz}, \tag{26}$$

to be solved for the three functions Λ, a, and b, where the function $K(u)$ is defined as

$$K(u) = 4 \int_0^\infty dp\, p F\left(u e^{-p^2}\right). \tag{27}$$

Equation (24) is nothing but a statement of the fact that during the EM beam evolution its power is conserved,

$$N = \pi \Lambda^2 a^2 = \pi \Lambda_0^2 a_0^2, \tag{28}$$

where Λ_0 and a_0 are respectively the initial amplitude and the initial "radius" of the EM beam at $z = 0$.

Using Eq. (24), the integration of Eq. (25) leads to

$$\frac{1}{2}\left(\frac{da}{dz}\right)^2 + V(a) = H = V(a_0), \tag{29}$$

where

$$V(a) = \frac{2}{a^2} - \frac{2a^2}{a_0^2 \Lambda_0^2} K\left(\frac{\Lambda_0^2 a_0^2}{a^2}\right) \tag{30}$$

plays the role of an effective potential for the evolution of the radius a. We have assumed that the initial beam have a plane front (or zero curvature) $[da/dz|_{z=0} = 0 = b(0)]$. Using the analogy with a particle in a potential well, we can acquire a deeper physical understanding of light beam dynamics. Choosing the initial beam radius a_0 to be equal to the equilibrium radius a_e, a stationary solution of Eq. (29) is obtained if $\partial V/\partial a|_{a=a_e} = 0$. Note that $-\partial V/\partial a$ is equal to the right-hand side of Eq. (25). The equilibrium radius of the beam is readily found to be

$$a_e^2 = 2 \left[K'(\Lambda_0^2) - \frac{K(\Lambda_0^2)}{\Lambda_0^2} \right]^{-1}. \tag{31}$$

In the subsequent analysis we will consider the small temperature case $T_\infty \ll 1$, and the case of moderately high temperature $T_\infty \sim 1$. We do not consider here the ultrarelativistic temperature case ($T_\infty \gg 1$) because applied model equations fail to adequately describe the plasma dynamics due to the neglect of heavy particle production.

In the small temperature case the nonlinear term in Eq. (18) reduces to a simple analytic expression

$$\Psi\left(|A|^2\right) = 1 - \left(1 - |A|^2\right)^{3/2}, \tag{32}$$

which gives the function $F(u)$ as

$$F(u) = u + \frac{2}{5}(1-u)^{5/2} - \frac{2}{5}. \tag{33}$$

We will apply the established general formalism to the saturating nonlinearity given by Eq. (32). The function K becomes [see Eq. (27)]

$$K(u) = 2u + \frac{4}{5}\left[(1-u)^{1/2}\left(\frac{46}{15} - \frac{22u}{15} + \frac{2u^2}{5}\right)\right]$$
$$- \frac{8}{5}\arctan(1-u)^{1/2} - \frac{4}{5}\log u - \frac{4}{5}\left(\frac{46}{15} - \log 4\right). \tag{34}$$

Note that normalized strength of the field $A\ (= A/A_c)$ is restricted from above $|A| \leq 1$. Above this value the wave-breaking of the field takes place. Substituting Eq. (34) into Eq. (25), we can investigate the nonlinear dynamics of the beam for different initial conditions.

Using Eq. (34) we can find the effective potential $V(a)$ [see Eq. (30)]. The shape of effective potentials in the case of $N = 8.79$ and $N = 9.74$ are illustrated in Fig. 2. From the shape of the potential we conclude that the "effective particle" (i.e. the beam) can be trapped in the potential well. The endpoints of these lines (for small a) correspond to the occurrence of zero density. As a decreases the beam amplitude Λ increases and approaches the unity at endpoints. Note that the potential at the endpoint is negative for $N = 9.74$. Thus, the cavitation can take place when the initial radius a_0 is large enough. It is obvious that $a(z)$ is an oscillating function provided that a_0 is in the trapping region.

The beam never diffracts if $H < 0$ which coincides with the general criterion established by Zakharov et al. [23]. Figure 3 shows the parameter regime where the solitary wave could be trapped. The central line denotes the equilibrium curve, and the shaded region between two dashed curves denotes a region where EM beam is trapped. The lower

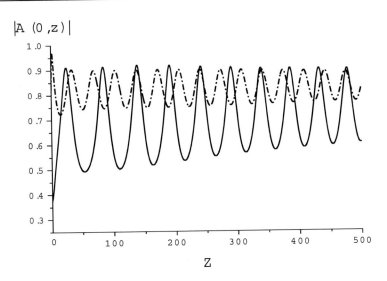

Figure 4: The beams parameters are oscillating around the equilibrium with initial focusing $(\Lambda_0 = 0.37)$ or defocusing $(\Lambda_0 = 0.95)$.

dashed line corresponds to the zero Hamiltonian. Thus, below this line, the electromagnetic beam with any parameters will diffract. Above the upper dashed line, the electromagnetic beam amplitude will grow up to wave-breaking limit. Therefore, the beam will be trapped in oscillatory regime provided that its parameters are situated in shaded region in Fig. 3.

4   Numerical Simulation

In the preceding analysis, we applied a variational approach involving a Gaussian trial function. The main limitation of this approach is that it is valid only in the aberration-less approximation, i.e. the approach is unable to account for structural changes in the beam shape. Such aspects of the beam dynamics are better delineated by numerical simulations. The detailed dynamics of arbitrary field distribution must be studied by direct simulations of Eq. (18). The guidelines for simulation are still provided by approximative analytical approaches.

The initial profile of the beam is taken to be Gaussian $A = \Lambda_0 \exp(-r^2/2a_0^2)$. In the case of small temperature $T_\infty = 0.01$ the initial parameters are taken in the trapping region shown in Fig. 3. The initial focusing or defocusing before reaching the equilibrium, can be seen in Fig. 4, where the z-propagations of the beam fields $|A(r = 0, z)|$ with same initial power $N = 9.5$ but with different amplitudes $\Lambda_0 = 0.37$ and $\Lambda_0 = 0.95$ are simultaneously drawn. Corresponding initial states of the beams are respectively situated on the left and right side of equilibrium curve in Fig. 3. Figure 4 shows that the beams parameters are oscillating around the equilibrium with initial focusing $(\Lambda_0 = 0.37)$ or defocusing $(\Lambda_0 = 0.95)$ in agreement with the prediction of variational approach.

Figure 5 shows the field distribution versus radius r and the propagating coordinate z for $\Lambda_0 = 0.37$. Note, however, that due to the appearance of the radiation spectrum the amplitudes of the field oscillations are monotonically decreasing with increasing z. For

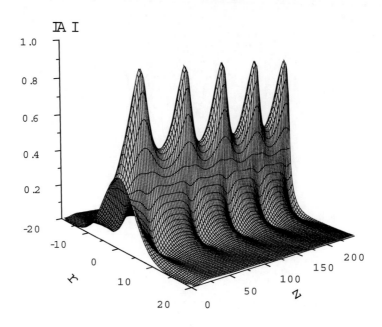

Figure 5: The field distribution versus r and z for $\Lambda_0 = 0.37$.

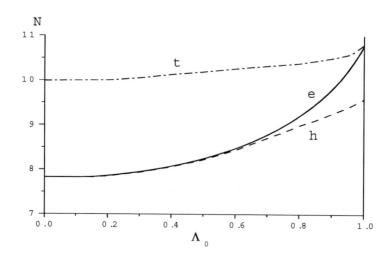

Figure 6: The trapping region in (N, Λ_0) plane obtained by numerical simulations for $T_\infty \ll 1$. The curve 'e' corresponds to the equilibrium state, the curve 'h' the zero Hamiltonian, and the curve 't' the trapping boundary, respectively.

larger z the formation of a ground solitonic state may take place due to the damp-out of the oscillations. If the initial profile of the beam is close to the equilibrium one, then the beam quickly reaches the profile of ground-state equilibrium, and propagates for a long distance without much distortion of its shape. Since variational approach is unable to account the structural changes in the beam shape and corresponding formation of radiation spectra we can expect that the trapping region in (N, Λ_0) plane obtained by numerical simulations will

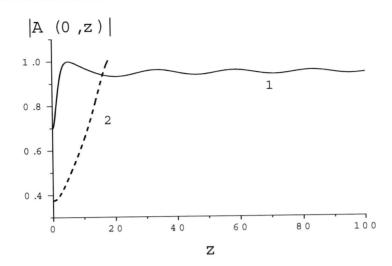

Figure 7: The field $|A(r = 0)|$ versus z for the cases when the beams parameters are in below (the curve 1) and above (the curve 2) the line t.

be different from the one shown in Fig. 3. The result of these simulations is presented in Fig. 6. The curve 'e' corresponds to the equilibrium state, the curve 'h' the zero Hamiltonian, and the curve 't' the trapping boundary, respectively. The Gaussian beam with initial parameters (i.e. N and Λ_0) in the region below line h has a positive Hamiltonian and will be diffracted. The trapping region of the beam is the area between lines h and t. The beam with the parameters in this region will be trapped in self-guiding regime of propagation and will either focus or defocus to the ground state, exhibiting damped oscillations around it. The initial focusing (defocusing) takes place if the parameters are in the region between lines t and e (between e and h). Note that the area of the trapping region shown in Fig. 6 is larger than it follows from variational approach (see Fig. 3). This enlargement of the trapping area is related to the radiation losses during the beam convergence to the equilibrium state.

The beam with parameters in region above trapping line t will focus until wave-breaking and plasma cavitation takes place. Figure 7 shows the plot $|A(r = 0)|$ versus z for the cases when the beams parameters are in the regions below (the curve 1) and above (the curve 2) the line t. One can see that in the first case the beam trapping takes place while, in the second case the beam amplitude increases up to the wave-breaking limit ($A = 1$).

For relativistic high temperature case the EM beam dynamics is similar to what we observed in the low temperature case. Figure 8 shows the trapping region for $T_\infty = 0.3$ (the notations are same as in Fig. 6). Quantitative difference stems from the fact that in the high temperature case the mass of the e-p pairs are modified by the temperature dependent G-factor. It is interesting to mention that in this case in the region of field localization the plasma temperature can be decreased considerably. In Fig. 9 we demonstrate $T(r)$ versus z during the trapped beam evolution toward the stable equilibrium.

It is thus confirmed by numerical simulations that the beam trapping is robust especially for low background temperature regime. The beam can achieve the equilibrium state from a wide range of parameters, even far from the equilibrium, as it is predicted by our analytic approach. However, the area of trapping region reduces as temperature increases.

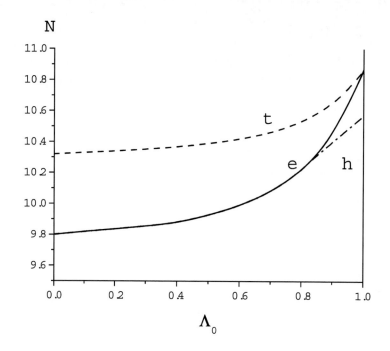

Figure 8: The trapping region in (N, Λ_0) plane for $T_\infty = 0.3$ (the notations are same as in Fig. 6).

5 Summary

We have investigated the nonlinear propagation of strong 2D EM radiation in a relativistic, unmagnetized electron-positron plasma. The treatment is fully relativistic — in the thermal motion as well as in the coherent motion of the plasma particles. The fact that relativistically hot e-p plasmas are capable of sustaining high amplitude localized structures of high amplitude electromagnetic fields should be important to understand the complex radiative properties of different astrophysical objects where such plasmas are considered to exist.

By applying a variational technique with a Gaussian trial function, we have demonstrated the possibility of different regimes of EM beam propagation. In particular, the trapped beam exhibits an oscillatory behavior around the stationary solution. However, since the system contains an upper bound of the field amplitude, the parameter regime is restricted for these oscillations. If the initial beam intensity is large enough, the ponderomotive force acts strongly leading to the beam over-focusing, and eventually the EM field gives rise to the plasma cavitation and wave-breaking.

The region in parameter space where the beam can be regularly trapped and oscillate is found. We have also performed the numerical simulations and demonstrated the parameter regime where beam trapping takes place. Comparing the results of variational approach and the results of numerical simulations we have confirmed that analytical estimate gives a qualitatively good prediction for nonlinear wave dynamics. Due to the radiation and self-deformation of the beam, however, it is shown that the beam may be more robust than it follows from variational method.

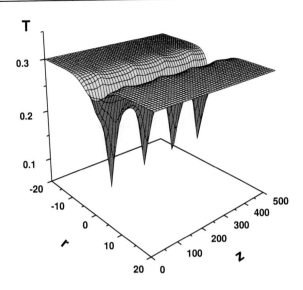

Figure 9: The plasma temperature $T(r)$ versus z during the trapped beam evolution toward the stable equilibrium.

We may apply the present results to the problem of gamma-ray bursts in the pulsar magnetosphere. Lyutikov and Blackman address the fundamental but unresolved issue — how Poynting flux is converted into gamma-rays and the subsequent emission characteristics [18]. Poynting-flux driven outflows from magnetized rotators are a plausible explanation for gamma-ray burst engines. These authors argue that, in a region near the rotation axis, the Poynting flux drives the large-amplitude electromagnetic waves "break" at radii $r_t \sim 10^{14}$ cm, and consequently the MHD approximation becomes invalid. In the "foaming" (i.e., relativistically re-connecting) regions formed during the wave breaking, the random electric fields stochastically accelerate particles to ultra-relativistic energies which then generate turbulent electromagnetic fields. Thus, the claim is that due to wave breaking such a radiation can be considered as a source of gamma-ray bursts. Our results indicate that the wave-breaking can take place at a distance much shorter than r_t.

The work of VIB and SVM has been supported by ISTC Grant G663.

References

[1] P.A. Sturrock, *Astrophys. J.* **164**, 529 (1971); M.A. Ruderman and P.G. Sutherland, *ibid.* **196**, 51 (1995); F.C. Michel, *Theory of Neutron Star Magnetospheres* (University of Chicago Press, Chicago, 1991).

[2] M.C. Begelman, R.D. Blandford, and M.J. Rees, *Rev. Mod. Phys.* **56**, 255 (1984).

[3] M.L. Burns, in *Positron-Electron Pairs in Astrophysics* edited by M.L. Burns, A.K. Harding, and R. Ramaty (American Institute of Physics, New York, 1983).

[4] S. Weinberg, *Gravitation and Cosmology* (Wiley, New York, 1972); P.J.E. Peebles, *Principles of Physical Cosmology* (Princeton University Press, Princeton, 1993).

[5] M. Perry and G. Mourou, *Science* **264**, 917 (1994); S.P. Hatchett *et al.*, *Phys. Plasmas* **7**, 2076 (2000).

[6] A.C.L. Chian and C.F. Kennel, *Astrophys. Space Sci.* **97**, 9 (1983).

[7] R.E. Kates and D.J. Kaup, *J. Plasma Phys.* **41**, 507 (1989).

[8] R.T. Gangadhara, V. Krishan and P.K. Shukla, *Mon. Not. Roy. Astron. Soc.* **262**, 151 (1993).

[9] T. Tajima and T. Taniuti, *Phys. Rev. A* **42**, 3587 (1990); K. Holcomb and T. Tajima, *Phys. Rev. D* **40**, 3909 (1989).

[10] V.I. Berezhiani and S.M. Mahajan, *Phys. Rev. Lett.* **73**, 1110 (1994); V.I. Berezhiani and S.M. Mahajan, *Phys. Rev. E* **52**, 1968 (1995).

[11] T. Tatsuno, V.I. Berezhiani, and S.M. Mahajan, *Phys. Rev. E* **63**, 046403 (2001); T. Tatsuno, V.I. Berezhiani, M. Pekker, and S.M. Mahajan, *Phys. Rev. E* **68**, 016409 (2003).

[12] M.J. Rees, *preprint* (http://xxx.lanl.gov/abs/astro-ph/9701162) (1997).

[13] V.V. Usov, *Nature* **357**, 472 (1992).

[14] C. Thompson, *Mon. Not. Roy. Astron. Soc.* **270**, 480 (1994).

[15] E.G. Blackman, L. Yi, and G.B. Field, *Astrophys. J.* **473**, L79 (1996).

[16] P. Mészáros and M.J. Rees, *Astrophys. J.* **482**, L29 (1997).

[17] W. Kluzniak and M.A. Ruderman, *Astrophys. J. Lett.* **505**, 113 (1998).

[18] M. Lyutikov, E.G. Blackman, *Mon. Not. Roy. Astron. Soc.* **321**, 177 (2001).

[19] V.I. Berezhiani, S.M. Mahajan, Z. Yoshida, and M. Ohhashi, *Phys. Rev. E* **65**, 047402 (2002).

[20] G.Z. Sun, E. Ott, Y.C. Lee, and P. Guzdar, *Phys. Fluids* **30**, 526 (1987).

[21] N.G. Vakhitov and A.A. Kolokolov, *Izv. Vuzov Radio Fiz.* **16**, 1020 (1973) [*Sov. Radiophys.* **9**, 262 (1973)].

[22] V. Skarka, V.I. Berezhiani, and R. Miklaszewski, *Phys. Rev. E* **56**, 1080 (1997).

[23] V.E. Zakharov, V.V. Sobolev, and V.C. Synakh, *Zh. Eksp. Teor. Fiz.* **60**, 136 (1971) [*Sov. Phys. JETP* **33**, 77 (1971)].

In: Advances in Plasma Physics Research, Volume 7
Editor: Francois Gerard

ISBN: 978-1-61122-983-7
© 2011 Nova Science Publishers, Inc.

Chapter 7

SUPERSONIC MOLECULAR BEAM INJECTION IN FUSION PLASMA

Lianghua Yao[*]

Southwestern Institute of Physics, P. O. Box 432, Chengdu 610041, China

ABSTRACT

There are three conventional techniques used to fuel fusion devices: gas puffing, ice pellet injection and neutral beam injection. Gas puffing is the simplest fuelling tool and it is generally used in all devices to establish primary plasma and control the plasma density by feedback, but the fuelling efficiency is quite low, in the range of 5-25 %. On the other hand, a pellet injected from inside the magnetic axis from the inner wall leads to stronger central mass deposition and thus yields deeper and more efficient fuelling. However, this injection system meets complex problems related to producing and launching as well as transporting ice pellets. The main goal of neutral beam injection is the heating of the plasma, but not fuelling the large fusion plasmas.

Supersonic molecular beam injection (SMBI) was first proposed and demonstrated on the HL-1 tokamak, was successfully developed and used on the HL-1M tokamak, and was then applied on the HT-7 superconducting tokamak, the Tore Supra superconducting tokamak, the W7-AS stellarator and the ASDEX Upgrade tokamak. SMBI can enhance the penetration depth and fuelling efficiency in the previous devices. With the new fuelling method, high densities of 8.2×10^{19} m^{-3} and 6×10^{19} m^{-3} were obtained for HL-1M and HT-7, respectively. A stair-shaped density increment was obtained with high-pressure multi-pulse SMBI that was just like the density evolution behavior during multi-pellets injection. A pneumatic pulsed SMB injector was developed in CEA/DSM/DRFC at Cadarache, which can work in presence of strong magnetic field and greatly increase fuelling efficiency in Tore Supra for 3 times than that of conventional gas puff.

Considering the relatively high temperature of edge plasma for the large tokamak with divertor configuration compared with the limiter one , a cluster jet injection (CJI), which is like the micro-pellet injection, will be beneficial to deeper injection and higher fuelling efficiency for both of the SMBI and gas puffing. The experiment on cryogenically cooled high-pressure hydrogen cluster jet injection into the HL-2A plasma

[*] E-mail address: yaolh@swip.ac.cn, Tel: 86-28-82850311, Fax: 86-28-82850300

was carried out and the fuelling effects were distinctly better than that of the room temperature one. This technique may be a candidate for the fuelling of the International Thermonuclear Experimental Reactor (ITER).

1. INTRODUCTION

The control of plasma density is an issue that continues to pose great problems [1]. After a long period (more than thirty years) of study, the ion temperature in tokamaks has now reached 40 keV or higher. The expected ignition ion temperature has been reached and exceeded, and the achieved temperature is about two orders of magnitude greater than that of the original fusion devices. In contrast to this situation, one cannot be satisfied with the progress of the study of plasma density. The available density in large tokamaks hardly exceeds $1 \times 10^{20} m^{-3}$, except in the case of a strong field. Looking back on the history of tokamak progress, one sees that the plasma density in present large devices is only two or three times higher than that of the earliest devices. This situation means that many difficulties will remain for a fusion reactor.

There are three conventional techniques used to fuel fusion devices: gas puffing (GP), ice pellet injection (PI) and neutral beam injection (NBI). A piezoelectric valve is commonly used to puff gas, it is the simplest fuelling tool and it is generally used in all devices to establish primary plasma and control the plasma density by feedback. Owing to lack of directionality from a puff of gas, most of the injected particles are absorbed at the wall surface of the vacuum vessel and then become recycling particles or are exhausted straight by the pump system. This fuelling method can only puff the gas particles at the plasma edge, so the fuelling efficiency is quite low, in the range of 5-25 % for different background plasma densities and different machines. On the other hand, pellets protected by their ablating particle cloud can penetrate deeply into the core of the plasma, resulting in density peaking and high fuelling efficiency as well as improved energy confinement. Especially a pellet injected from inside the magnetic axis from the inner wall leads to stronger central mass deposition than a pellet injected from the outside mid-plane, and thus yields deeper and more efficient fuelling [2]. However, to perform a more efficient operation of multi-pellet fuelling with high field side injection, this injection system meets more and more complex problems related to producing and launching as well as transporting ice pellets. All these troubles limit this fuelling method as a routine tool. The main goal of NBI is the heating of the plasma, but it is also an efficient fuelling tool for some devices, such as the W7-AS stellarator [3]. Because the ratio of the amount of injected particles to the deposited energy of the particles is small for high power neutral beam, it is difficult to use high-energy neutral beam injection to fuel large fusion plasmas.

In order to reform the gas fuelling technique, a new fuelling method, supersonic molecular beam injection (SMBI) was first proposed and demonstrated on the HL-1 tokamak [4], was successfully developed and used in its modification device HL-1M [5,6,7,8], and was then applied on the HT-7 superconducting tokamak [9], the Tore Supra superconducting tokamak [10], the W7-AS stellarator [11], and the ASDEX Upgrade tokamak [12]. With the new fuelling method, high densities of $8.2 \times 10^{19} m^{-3}$ [13] and $6 \times 10^{19} m^{-3}$ [9] were obtained for HL-1M and HT-7, respectively. SMBI can enhance the penetration depth and fuelling efficiency, as well as to reduce both the injected particle wall-surface interaction and the impurity content in the plasma. It is a significant improvement over conventional gas puffing. A stair-shaped density increment was obtained with high-pressure multi-pulse SMBI that was just like the density evolution behavior during multi-pellets injection. Within the SMB the clusters may play an important role during the injection process. There is a mechanism for interpreting the deeper penetration of the SMB, a simple collective model based on the high flux of neutral particles with a very high density of 4.0×10^{24} m^{-3}.

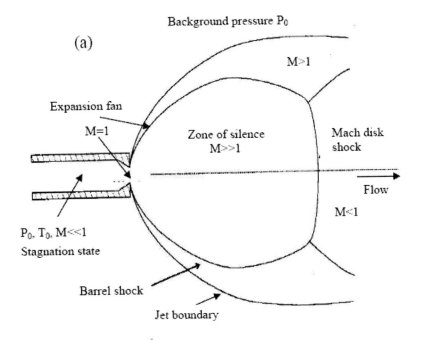

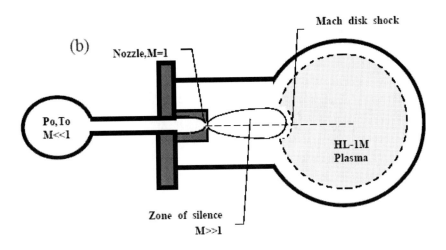

Figure 1. (a) Continuum free jet expansion. (b) Schematic diagram of supersonic molecular beam injection into the HL-1M plasma.

The formation and temporal evolution of a jet of hydrogen or deuterium clusters produced in the supersonic expansion into vacuum were studied via Rayleigh scattering under the codition of the gas at nitrogen temperature and moderate backing pressure. The cluster size or the paticle number within the cluster can be estimated.

Recently the experiment on cryogenically cooled high-pressure hydrogen cluster jet injection into the HL-2A plasma was carried out, and the particle injection depth and the fuelling efficiency were distinctly better than that of the room temperature one.

The alternative injection mode is to vary the injection location from the low magnetic field side to the high field side, this is an attempt for the benefit of drift effect along the toroidal field gradient. A pneumatic pulsed SMBI was developed in CEA/DSM/DRFC at Cadarache, which can work in presence of strong magnetic field and shorten the distance between the nozzle of the injector and the edge plasma, and greatly increase fuelling efficiency in Tore Supra for 3 times than that of conventional gas puff, and the total quantity injected was 30 % lower than for the later fuelling method.

2. CHARACTERISTIC PROPERTIES OF SMB AND CLUSTER

The supersonic molecular beam source used here is in fact a free jet. Figure 1 shows the features of a free jet expansion under continuum conditions and its injection into the HL-1M plasma. The source consists of a small chamber with a short converging nozzle, and the working gas can be kept at a definite pressure P_0 and temperature T_0 in the chamber. The gas, with a small start velocity (Mach number M<<1), is accelerated by an imposed pressure difference $(P_0 - P_b)$ through the nozzle to enter into the vacuum chamber. If the background pressure P_b of HL-1M is low enough and the ratio P_0 / P_b exceeds a critical value

$$G \equiv ((\gamma+1)/2)^{\gamma/(\gamma-1)} \quad (1)$$

which is less than 2.1, then M = 1 at the throat of the nozzle, where $\gamma = c_p/c_v$ is the heat capacity ratio ($\gamma = 5/3$ for a monoatomic gas and $\gamma = 1.4$ for a diatomic gas). Beyond the exit the gas flow expands isentropically. As the flow area increases, the supersonic flow velocity increases, and the Mach number continues to increase and becomes far greater than 1 in the zone of silence [14]. The characteristic dimension of the supersonic area X_M, measured in terms of nozzle diameter d, is given by [15]

$$X_M/d = 0.65 \ (P_0/P_b)^{1/2}. \quad (2)$$

For a given pumping speed $\phi(m^3/s)$, this gives a constant Mach disk position X_M, which varies as $T_0^{1/4}$ of gas source. For a continuum free jet of hydrogen, the final result is

$$X_M = 55(T_0/300K)^{1/4}(\phi(m^3/s))^{1/2} mm \quad (3)$$

It is seen from Eq. (3) that the increase of the dimension of supersonic area X_M is severe for pumping speed. In the present experiment the shortly pulsed operation of SMB was used, so the X_M as long as 40 cm is feasibly reached and the requirements for pumping system are significantly reduced.

In the initial collision-dominated regime the expansion is isentropic with flow velocity increasing up to its limit [16]

$$ v_{\lim it} = \left(\frac{2}{\gamma-1}\right)^{1/2} \left(\frac{\gamma k T_0}{m}\right)^{1/2} = \left(\frac{\gamma}{\gamma-1}\right)^{1/2} \left(\frac{2kT_0}{m}\right)^{1/2} \qquad (4) $$

where k is the Boltzmann constant, m the molecule (atom) mass and T_0 the Kelvin temperature in the gas source. As shown in Eq. (4), the limit velocity of SMB is equal to $[2/(\gamma-1)]^{1/2}$ times as large as the sonic velocity or $[\gamma/(\gamma-1)]^{1/2}$ times the maximum probability velocity.

In the HL-1M experiment, $d = 0.1$ mm and $P_0/P_b \approx 5\times10^7$, so $X_M \approx 47$ cm, which is longer than the distance 40 cm from the nozzle to the edge plasma of the HL-1M.

Pulsed SMB has high instantaneous intensity, high speed, small spread of velocity with narrow angular distribution and low gas consumption. Gas pulses of tens of milliseconds are commonly used in the experiments to control the edge recycling and to satisfy the general requirements of HL-1M gas fuelling.

The quantity of gas passing through the nozzle is calculated from the equation [17]

$$ j = \left(\frac{\gamma R T_0}{\mu}\right)^{1/2} \left(\frac{2}{\gamma+1}\right)^{\frac{\gamma+1}{2(\gamma-1)}} n_0 A \text{ particles / s,} \qquad (5) $$

where R is the molar gas constant, 8.314 J / mol·K, n_0 is the gas particle number per cubic metre in the gas source, A is the cross-section of the nozzle diameter and μ is the molecular molar mass.

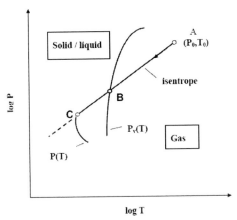

Figure 2. Schematic of cluster formation in free jets **B_** saturation point; **C_** onset of condensation point.

For real gases (say, at point A) the adiabatic expansion of the free jet crosses the steep vapor pressure curve as shown in Figure 2, and the gas becomes supersaturated at some point B. Onset of condensation (point C) depends on both the thermodynamic state given by the saturation point B, and the kinetics and time scale of the expansion, determined by nozzle size d and source state A (gas pressure and temperature). There is still no general theory to predict the onset point and thus the formation and growth of clusters in the free jet. However, the onset of clustering and size of clusters produced can be described by an empirical scaling parameter Γ^* referred to as the Hagena parameter [18].

$$\Gamma^* = k \frac{(d/\tan\alpha)^{0.85}}{T_0^{2.29}} P_0 \qquad (6)$$

where d is the nozzle diameter (μ m), α the expansion half angle ($\alpha = 45°$ for sonic nozzle, $\alpha < 45°$ for supersonic), P_0 the pressure (mbar) behind the valve, T_0 the pre-expansion temperature (Kelvin), and k a constant related to bond formation. Clustering generally begins for $\Gamma^* > 100\text{-}300$, with the number of atoms per cluster N_c, scaling as $N_c \propto \Gamma^{*\,2.0\text{-}2.5}$.

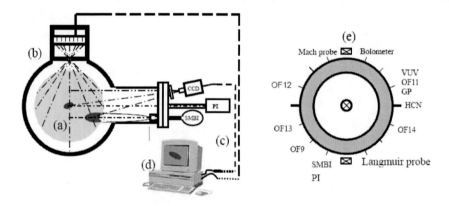

Figure 3. Experimental setup of PI & SMBI in the HL-1M tokamak. (a) HL-1M plasma, (b) 20 channel PIN array, (c) 40m long optical fibre, (d) Computer data A&P system, (e) HL-1M plasma tours and diagnostics around the torus.

3. SMBI IN HL-1M

3.1   Experimental Setup

HL-1M has a major radius R = 1.02 m, minor radius a = 0.26 m, $B_t \leq 3$ T and $I_p \leq 300$ kA, with two full graphite limiters located at toroidally symmetrical sections. The experimental set-up of SMBI in HL-1M is shown in Figure 3. A CCD camera was mounted 8.7 cm above the midplane and at an angle of 13.4° to the PI line, which is along a major radius of the midplane. The SMBI line is 9 cm below and parallel to the PI line. A detector array for H$^\alpha$ emission intensity which includes 20 PIN diode channels is located on the top port, on the same plasma cross-section as the PI and SMBI lines. The detector system consists of the collimating hole, interference filter, detectors and preamplifiers, and makes up a pinhole camera. The center wavelength of the interference filter is 656.3 nm. In order to determine the pellet penetration depth, we made an assumption that the instant position of the pellet coincides with that of the H_α emission peak at a given time. The sights of the array cover the whole plasma cross-section. The H_α emissivity distribution along the injection line was obtained with asymmetric Abel inversion of measured 20-channel H_α emission intensity. A series of diagnostics were installed around the torus. H_α signals from edge plasma were used to monitor the particles transport behavior of the particles in the various toroidal sectors during beam injection. The effects of SMBI on the edge plasma temperature and density were detected by Langmuir probes movable in radial direction. Single and six channel HCN interferometers were used to measure the line averaged central electron density and the density profile, respectively. The vacuum ultra violet (VUV) spectrometer is a useful tool for finding the impurities in the beam and the features of the doping gas. A 2 mm microwave scanning heterodyne receiver (ECE), which has been calibrated by a noise source and soft X-ray spectrometer, was used to measure the electron temperature profile under different discharge conditions. The plasma energy confinement time τ_E is measured by diamagnetic techniques.

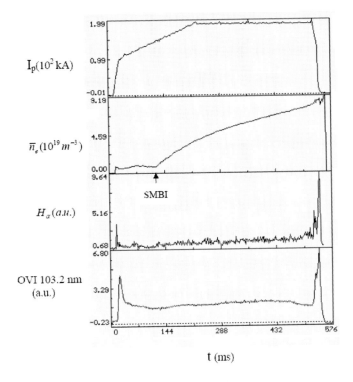

$$t \ (ms)$$

Figure 4. SMBI is a clean and high efficiency fuelling method The impurity content (Oxygen VI) approximately remains constant while the electron density increases for 7 times. Ohmic discharge: Shot 4965, I_p = 190 kA, B_t = 2.4 T.

3.2   High Density

With the high density operation in the fuelling experiments, different wall conditioning techniques (boronization, siliconization and lithium coating [19,20]) and fuelling methods (GP, SMBI and PI [21,22]) have been used in HL-1M. The best way to obtain the maximum density is SMBI fuelling with slow current ramp-up after siliconization of the first wall. The highest densities on HL-1M are 5.5×10^{19} m^{-3} for PI, 7.0×10^{19} m^{-3} for GP and 8.2×10^{19} m^{-3} for SMBI, with density increase rates dn$_e$/dt of $\geq 1.5 \times 10^{22}$ m^{-3}s^{-1} ; $\leq 3.8 \times 10^{20}$ m^{-3}s^{-1} and $\leq$ 1.5×10^{21} m^{-3}s^{-1}, respectively [8,13]. The experiments proved that the impurities and recycling from the first wall play an important role in achieving a density limit. Improved conditioning of the vessel walls will lead to a higher density. The higher density peaking factor Q_n after SMBI is of great benefit to obtaining high density. The maximum density obtained was $\bar{n}_e = 8.2 \times 10^{19}$ m^{-3} (shot 4965, pure ohmic heating hydrogen plasma with current slowly rising to 186 kA, $B_t = 2.4$T), as shown in Figure 4; this was about 80% of the Greenwald density limit.. The highest density limit is a factor of 1.4 of the Greenwald limit for a low current $I_p = 120$kA plasma [13]. The density limit in deuterium is somewhat higher than that in hydrogen, and the isotope effect scaling is $\bar{n}_e \propto m^{1/3}$. The Hugill diagram for a HL-1M plasma with GP, SMBI and PI was shown in Figure 5. The maximum Murakami constant is $C_M = 3.5 \times 10^{19}$ m^{-2} T^{-1} for Ohmic discharge with SMBI.

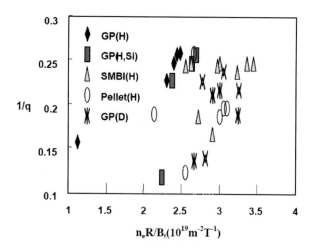

Figure 5. Hugill plot of density limit results for the ohmic discharges in HL-1M GP(H)_ gas puffing in hydrogen, GP(H,Si)_ gas puffing in hydrogen after siliconization, MBI(H)_ supersonic molecular beam injection in hydrogen, Pellet(H)_ hydrogen pellet injection, GP(D)_ gas puffing in deuterium.

3.3   Impurity Problem of SMBI

Some properties of the impurities in HL-1 and HL-1M are studied by use of the VUV spectroscopic diagnostic under different discharge conditions [23]. One of the experimental results shows that the impurity oxygen ion (O VI, 103.2 nm) spectroscopic intensity varies directly as the central line averaged electron density $\overline{n}_e$ at the steady state. The prerequisite for the regularity is that the electron temperature profiles remain constant over the given duration. However, the previously observed regularity disappears in the HL-1M plasma with SMBI fuelling; even if the density increases by a few times, the spectroscopic intensity of oxygen ion remains approximately constant. This is caused by the direct injection of the beam particles, which are cleaner than the recycling particles from the first wall. It is the lower fuelling efficiency for GP that results in the increment of recycling particles. On the other hand, the variation of intensity of the nitrogen ion (N IV, 99.5 nm) or oxygen ion (O VI, 103.2 nm) spectrum follows the variation of $\overline{n}_e$ during the doping gas SMBI with less impurity (air), as shown in Figure 6. The reason is that if the mixture consists of a small fraction of impurity (heavy) particles, as the mixed gas at pressure P_0 is accelerated by an imposed pressure difference ($P_0 - P_b$) through the nozzle to get into the vacuum chamber (pressure P_b), at the same time the gas particles are expanded as a result of the many collisions. The impurity particles will be accelerated to a speed v_i which in common with that of light particles [24]

$$v_i = \left(\frac{2\overline{\gamma}kT}{(\overline{\gamma}-1)\overline{m}} \right)^{1/2} \qquad (7)$$

where $\overline{m}$ is the mean molecular mass and $\overline{\gamma}$ is the mean specific heat ratio for the mixed gas. For a doping gas with a small fraction of impurities, $\overline{\gamma}$ and $\overline{m}$ approximately equal pure working gas specific heat ratio and molecular mass, respectively. The impurity velocity then becomes supersonic. Thus it may be concluded that SMBI fuelling would reduce the impurity accumulation during the period of density increase, and that it is also an effective method for impurity gas injection into the plasma.

3.4   Hollow Profile of T_e and Negative Magnetic Shear after SMBI

In the experiments with hydrogen SMBI plasmas, the hollow electron temperature profile phenomena sometimes appeared at the end of the plasma current plateau, where the plasma density had already reached a high level and was continuously peaking. A typical example is shot 4957: 12 hydrogen molecular beam pulses were injected from 50 ms to 550 ms, the line averaged electron density $\overline{n}_e$ reached 7.5×10^{19} m^{-3} at the plasma current plateau and the Q_n value increased quickly from 1.45 to 1.75 in the time range from 500 ms to 580 ms. An apparent concave T_e profile appears from 540 ms and then develops into a centrally hollow profile, as shown in Figure 7. Ohmic shear reversed configurations have been obtained in HL-1M by combined control of plasma current ramp-up and SMBI. They are characterized by hollow electron temperature profiles and peaked density profiles. An equilibrium reconstruction code which uses 32 magnetic probes has been applied to the HL-1M database for deriving the current density profile and the safety factor profile [25].

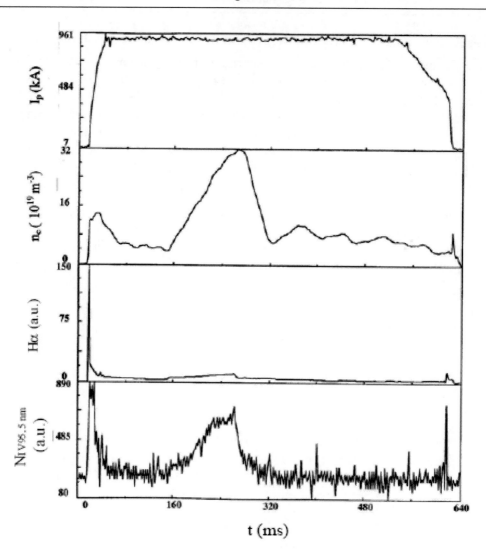

Figure 6. The effect of the doping gas SMBI with less impurity (air), the Nitrogen can be deeply injected into plasma.

3.5   High Performance of High-Pressure Multi-pulse SMBI

The recent beam injection experiments have been carried out by increasing the pressure of the hydrogen gas source from 0.5 MPa to over 1.0 MPa for enlarging the characteristic dimension of the supersonic area and further enhancing the beam injection depth, and leading to a higher number of particles injected. According to Equation (6), when the nozzle diameter $d = 100$, the pressure behind the valve $P_0 = 5000\text{-}11000$, $T_0 = 300$ and $k = 184$ and 1400 for hydrogen and oxygen, respectively. The Hagena parameter Γ^* is about 200 in the high-pressure SMBI experiments, so cluster onset may occur.

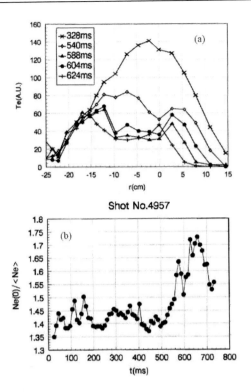

Figure 7. Variation of (a) T_e profile and (b) density peaking factor Q_n after hydrogen SMBI into the HL-1M plasma in shot 4975.

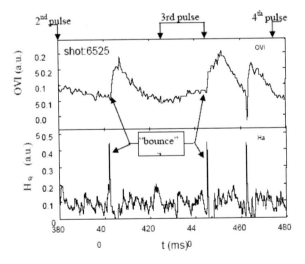

Figure 8. High intensity emission signals of H_α from the plasma and OVI (103.2nm) from the core plasma by VUV diagnostics during five pulse SMBI. One of the spikes appears during the 3rd beam pulse and two spikes appear in the interval of the 2nd pulse and the 3rd pulse, and of 3rd pulse and 4th pulse, respectively.

* The valve is made by Parker Hannifin Valve Corporation and specially designed to produce supersonic molecular beam.

Actually, there may be an empirical threshold of the pressure, P_0 = 1.0 MPa, for hydrogen clustering with the series 99 valve* in the HL-1M SMBI experiments. The consequence of increasing P_0 is apparent, when $P_0 \geq 1$ MPa, a few spikes of H_α emission appear during the beam pulse or in the interval between two beam pulses in the case of multi-pulse SMBI. The H_α emission signals with some high intensity spikes for multi-pulse SMBI are similar to the case of multi-pellet injection in HL-1M at the same injection cross-section, which are shown in Figure 5 and 2 of [26], respectively. A series of interesting features have been obtained during the interaction of the high-pressure SMB (including clusters) with the magnetically confined plasma.

The characteristic signals of the diagnostics, which occur nearly simultaneously, may show evidence of the onset of clustering, which resulted in valuable fuelling effects. On the other hand, the higher number of injected particles at higher injector pressure could also play a role. In shot 6525, there are five SMB pulses injected into the plasma at pressure P_0 = 1.0 MPa, the first pulse from 300 ms to 315 ms, the second from 365 ms to 380 ms, the third from 425 ms to 445 ms, the fourth pulse from 476 ms to 445 ms and the fifth from 530 ms to 550 ms. The spikes of H_α emission from the plasma and the corresponding spikes of the VUV spectrum (OVI 103.2 nm) have been detected, as shown in Figure 8. This means that within the SMB, hydrogen and oxygen particles nearly simultaneously enter into the centre of the plasma. Only in the beam, the residual gas oxygen has nearly the same velocity than the hydrogen, and oxygen clusters more easily than hydrogen. The penetration depth and injection speed of the high-pressure SMB in the HL-1M plasma were obtained from the contour plot of the evolution of H_α emission intensity based on measurements from the top view of the PIN diode array and Abel inversion, as shown in Figure 9. It is shown that the time at which the hydrogen particles penetrate into the core region of the plasma corresponds to the time at which the H_α spike appears. The penetration speed of the high pressure SMBI in the plasma is about 1200 m/s. This value corresponds with the diagnostic results from the CCD camera.

In comparison to low pressure example in shot 4957 at P_0 = 0.31 MPa (Figure 7), after the beams injection, the density peaking factor Q_n reached a maximum value of 1.75 and a hollow electron temperature profile appears as t = 540 ms. Due to the increase in the penetration depth of hydrogen particles, the rate of increase of electron density, $d\overline{n}_e / dt$, was up to 7.2×10^{20} m^{-3} s^{-1} without disruption, the H_α emission, which is measured by the detector array with low sensitivity located at the top of the cross-section for injection beam, appears without spikes. This means that the gas particles entering into the plasma core are rather few; a large proportion of the injected neutral particles is ionized in the edge of the plasma. In the start phase, the neutral particles in the beam concentrate mainly in the edge of plasma, and afterwards diffuse slowly into the deeper region of the plasma.

A stair-shaped density increment was obtained with high pressure multi-pulse SMBI that was just like the density evolution behavior during multi-pellet injection, as shown in Figure 10 of shot 7300. The plasma line averaged electron density increases from $0.77 \times 10^{19} m^{-3}$ to $5.67 \times 10^{19} m^{-3}$ during three-pulse SMBI. The first beam pulse indicated in Figure 10 starts from t = 150.1 ms to 163.1 ms, the second from t = 200.8 ms to 219.7 ms; they are performed during the period of current ramp up for increasing the plasma stored energy and seeking the posibility of negative shear, and the third from t = 246 ms to 266.5 ms.

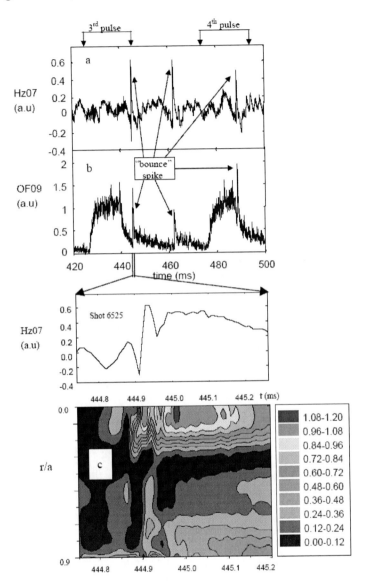

Figure 9. Variation of H_α emission from (a) the top view chord plasma and (b) the edge plasma, and (c) contour plot of H_α intensity during multi-pulse SMBI.

The gas pressure of the beam source used for this shot is 1.05 MPa. There is a superthermo-emission from the non-Maxwellian electron velocity distribution before t = 200 ms, and the plasma stored energy and ion temperature can not precisely be measured before t = 150 ms for low electron density, less than $1.0 \times 10^{19} m^{-3}$ even if the additional gas puffing starts from t = 60 ms. The hollow profiles of electron temperature obtained by ECE measurements during the second beam injection provide evidence that some of the particles have been injected straight into the center of the plasma (as shown in Figure 11) rather than after the beams injection, the injected cold particles gradually accumulate into the plasma centre and result in the decrement of core plasma electron temperature, as shown in Figure 7 for short 4957. During the period of the current ramp up the plasma density and stored energy are at low level until the first SMB pulse injections; then the plasma stored energy W sharply increases. After the three-pulse SMBI the plasma stored energy gradually increases and then maintains a stationary state (as shown in Figure 12). The ion temperature detected by a charge exchange neutral particle energy spectrometer increases only after the first beam pulse injection and then reachess a stationary value of 0.6 keV. The plasma energy confinement time measured by diamagnetism techniques was 30 ms in this shot. The increase of the plasma stored energy is mainly due to the increment of the plasma density during SMBI.

In addition, around the injection port, the edge D_α signals are a positive wave, which is consistent with that of injection beam pulses, whereas the D_α signals far from the injection port become negative pulses. A typical example of edge D_α signals is shown in here (Figure 13) of [27] obtained in W7-AS during and after SMBI with 3 MPa deuterium. This means that during the period of SMBI the recycling process is reduced at the wall surface. This reduced recycling demonstrated the effectiveness of high pressure SMBI as a promising fuelling tool for steady-state operation.

3.6   Edge effects induced by SMBI

A Mach/Langmuir six probes array assembly is mounted on a long shaft, which can be moved radially inboard and outboard and rotated around the axis of the shaft in the HL-1M torus [28]. In the assembly, three to six 6 pins act as a Reynolds stress probe and one to two pins act as collecting electrodes of a Mach probe to collect ion saturation currents in the poloidal direction. The probes were oriented with respect to the magnetic field direction to avoid shadows between them. The perpendicular component of the fluid velocity, $v_{i\perp}$, is tangential to the magnetic surfaces and points nearly in the poloidal direction. It can be written as [29]

$$v_{i\perp} = -\frac{1}{eZ_i n_i B}\frac{\partial P_i}{\partial r} + \frac{E_r}{B} = v_{diam} + v_{E\times B} \qquad (8)$$

where the first term on the right hand side of the equation is small compared to the second term in the present experiment; $v_{i\perp}$ is basically the $E\times B$ flow. We adopt alternative relation, the poloidal flow velocity v_{pol}, which is perpendicular component of the fluid velocity $v_{i\perp}$, is estimated by [30]

$$v_{pol} = \frac{J_{pol}}{n_e e}, \qquad (9)$$

where J_{pol} and n_e are poloidal current density and electron density, respectively. The present experimental data verify that $v_{pol} \geq E \times B$ flow, so that the formula (9) is applicable to the HL-1M experiments.

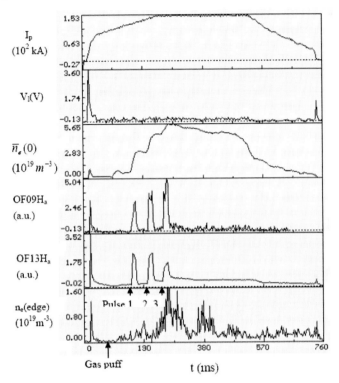

Figure 10. Plasma density variations during three- pulse SMBI in shot 7300. First pulse from 150.1 ms to 163.1 ms, second from 200.8 to 219.7 ms, third from 246.6 to 266.5 ms. OF09 and OF13 : edge H_α signals around the HL-1M torus; n_e(edge): edge electron density at r = 25.5 cm.

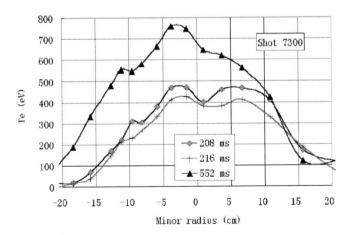

Figure 11. Variations of electron temperature profiles during and after three-pulse SMBI in shot 7300. First injection from 150.1 ms to 163.1 ms, 2nd injection from 200.8 ms to 219.7 ms, 3rd injection from 246.6 ms to 266.5 ms. There is a super-thermo emission before t = 200 ms.

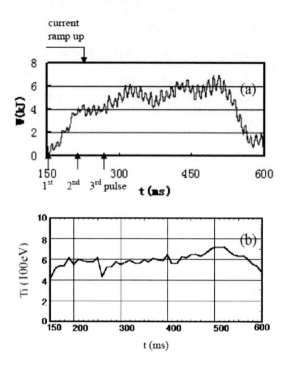

Figure 12. (a) Plasma stored energy W and (b) ion temperature T_i versus time in shot 7300 after three pulse SMBI.

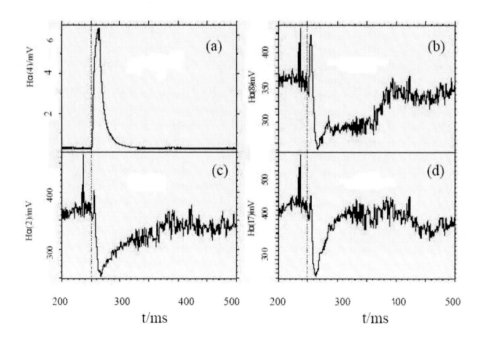

Figure 13. Edge plasma H_α time traces at different section of the W7-AS torus during and after SMB injection, beam injection time at t = 250 ms (dashed line). (a) Beam injection port; (b), (c) and (d) the other location far from the injection port.

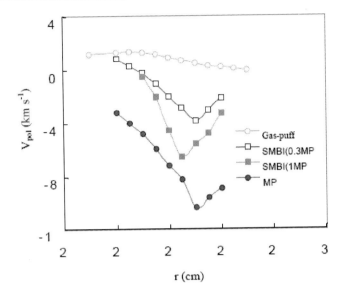

Figure 14. Radial profile of V_{pol} in the edge plasma during gas-puff, SMBI and MPI, respectively.

In the experiment of SMBI, the depth of SMB penetration into the plasma increases with the increase of the working gas pressure of the beam source. As the pressure increases from 0.3 MPa to 1 MPa, the peak value of the plasma poloidal velocity V_{pol} has an apparent increment from 3.8 km /s (at r = 25cm) to 6.5 km /s (at r = 24.5 cm) in the electron diamagnetic direction, as shown in Figure 14. The figure also reflects a comparison of V_{pol} for the three fuelling method, gas puff, SMBI and pellet injection, because the penetration depth of SMB is higher than that of gas puff and is less than that of pellet injection. When the gas pressure increases from 0.3 MPa to 1.0 MPa, the peak value of Reynolds stress moves from the location r = 25 cm to r = 24.5 cm. The results indicate that sheared poloidal flow can be generated in a tokamak plasma due to variation of the radial Reynolds stress with high-pressure SMBI. The increase of the poloidal rotation velocity shear may reduce the turbulent fluctuations, and the plasma confinement would be improved.

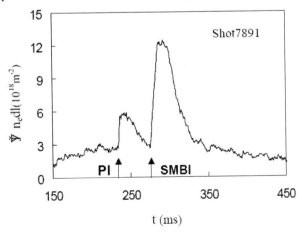

Figure 15. Comparison of density incre ase rate for PI and SMBI in shot 7891.

Ohmic discharge: I_p = 145 kA, B_t = 2.5 T. Pellet: $\phi\,1.2\times1.5$, flying velocity 1000 m / s, $d\bar{n}_e\,/\,dt = 1.3\times10^{21}\,m^{-3}s^{-1}$. SMB: Pulse duration 12 ms, hydrogen pressure 1.15 MPa, $d\bar{n}_e\,/\,dt = 1.5\times10^{21}\,m^{-3}s^{-1}$.

3.7   Comparison of Density Increase Rate for SMBI and PI

For a global comparison of the fuelling effects, the SMBI line is 9 cm below and parallel to the PI line, which is along a major radius of the mid-plane of HL-1M torus. Apart from the external injection particles, the fuelling sources must include recycling particles from the wall surface, which is strongly dependent on wall condition and is a temporal function. Sometimes the recycling particles are the main fuelling source, especially in high-density operation. Accurate calculation of fuelling efficiency requires the data on the total number of recycling particles, which is difficult to measure. In addition, the fuelling depth directly affects the fuelling efficiency and the density peaking factor. The fuelling efficiency is also a function of background plasma density. What is measurable is the increase rate of the central channel electron density, $d\bar{n}_e\,/\,dt$, which contains the fuelling ability and is directly proportional to the injection gas pressure or injected pellet size. We suggest that the maximum obtainable $d\bar{n}_e\,/\,dt$ value is used, instead of the fuelling efficiency, to study and compare the fuelling effects of SMBI and PI in identical Ohmic discharges.

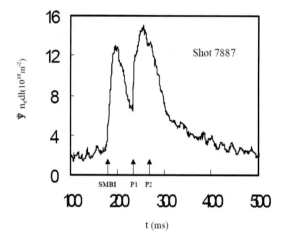

Figure 16. Comparison of density increase rate for single pulse SMBI and two-pellet injection in shot 7887.

Ohmic discharge: I_p = 145 kA, B_t = 2.5 T. Pellet: $\phi\,1.2\times1.5$, flying velocity 1000 m / sec, $d\bar{n}_e\,/\,dt = 1.3\times10^{21}\,m^{-3}s^{-1}$. SMB: Pulse duration 20 ms, hydrogen pressure 1.1 MPa, $d\bar{n}_e\,/\,dt = 8.7\times10^{20}\,m^{-3}s^{-1}$.

A comparison of fuelling efficiencies was made for high-pressure SMBI and small ice PI in identical Ohmic discharge during the period of plasma current plateau. The only difference in the conditions was the injection sequence. The increase in the rate of change of electron density, $d\bar{n}_e\,/\,dt$, for high-pressure SMBI was nearly equivalent to that of the small ice PI, as shown in shots 7891 and 7887 in Figures 15 and 16, respectively. The ice cylinder of the pellet was of 1.2 mm diameter and 1.5 mm height, and the average injection velocity of the ice cylinder before it entered the plasma was about 1000 m s^{-1}. The $d\bar{n}_e/dt$ values of the high-pressure SMBI and small PI for shot 7891 are 1.5×10^{21} m^{-3} s^{-1} and 1.2×10^{21} m^{-3} s^{-1}, respectively. For shot 7887, the $d\bar{n}_e/dt$ of SMBI is reduced to 8.7×10^{20} m^{-3} s^{-1}, whereas the value of $d\bar{n}_e/dt$ for PI is the same as that for shot 7891. The fact that the gas pressure of the SMB of shot 7891 is less than that for shot 7891 by 0.05 MPa leads to a decrease in the injected particle flux, which may be responsible for the decrease on the $d\bar{n}_e/dt$ value. Normally, $d\bar{n}_e/dt$ for the larger ice pellet, with a 1.5 mm cylinder diameter, is higher than that for the small one by an order of magnitude, reached 1.5×10^{22} m^{-3} s^{-1}, in HL-1M as described in [13,23].

4. CJI in HL-2A

Early in 1972 R. Klinglhofer and H. O. Moser presented a proposal [31] that intense hydrogen clusters with a velocity of about 500 m / s can be used for the refuelling of a thermonuclear plasma or for the compensation of particle losses in a thermonuclear device. They declared that the mean size of their hydrogen cluster reached 10 ^{9} atoms, but one did not find any record for using the cluster beam to fuel any machine. Up to 21 century, in 2005, the condensed molecular beam technique was developed in our institute and applied to injecting into the HL-2A plasma.

4.1   Experimental Setup

The main parameters of the HL-2A tokamak are R = 1.65 m, a = 0.4 m, B$_t$ = 2.8 T and I$_p$ = 0.48 MA [32]. The divertor of the machine is characterized with two closed divertor chambers, but now it is operated with lower single null configuration. The experimental set-up of SMBI system in HL-2A and the detail structure of the molecular beam valve with cooling trap are shown in figure 17. The distance between the nozzle of the valve and the edge plasma is about 1.28 m. The cylindrical valve body was surrounded by a close fitting liquid nitrogen reservoir for cooling the valve body and decreasing the working gas temperature. The hydrogen cluster jet used for the experiments is in fact a free jet. For real gases, the adiabatic expansion of gas through a nozzle into vacuum results in substantial cooling in the frame of the moving gas, and atoms or molecules that interact weakly at low temperature can form clusters as a result. Attractive forces between atoms can be hydrogen bonding, and the clustering effect is primarily determined by the temperature and pressure of the gas reservoir, shape and size of the nozzle, and strength of the inter-atomic bonds formed. There is no rigorous theory to predict cluster formation in a free jet expansion. However, the onset of clustering can be described by an empirical scaling parameter Γ^* referred to as the Hagena parameter, as shown as formula (6).

4.2 Rayleigh Scattering

The hydrogen clusters were produced by the adiabatic expansion of high-pressure (more than 0.5 MPa) hydrogen gas through a nozzle into vacuum and the temperature of the cluster source composed of the pulsed valve and the nozzle was cryogenically cooled to below 90K in this way. A Rayleigh scattering method was employed to measure the hydrogen cluster size. The layout of the Rayleigh scattering experiment was illustrated in figure 18 [33]. A 532 nm pulsed laser beam with energy of approximately 0.2 mJ was roughly focused by a lens and introduced into the vacuum chamber to intersect the hydrogen plume at right angles. The distance is 2 mm from the intersection to the valve nozzle. A photomultiplier, which was arranged in the direction perpendicular both to the cluster jet and to the incident laser beam, was used to amplify the scattered light from the hydrogen clusters. The signal from the photomultiplier was then fed to a digital oscilloscope (LeCroy 9350AL) for recording. The data were obtained after averaging over more than 20 shots for every datum. From the classical Rayleigh scattering theory and under the assumption that all the atoms or molecules of the stagnation gas (pressure P_0) are condensed into clusters in the expansion process, the intensity of the scattered light S_{RS} changes with the gas backing pressure P_0 and the average cluster size $\bar{N}_c$ (the average number of atoms per cluster) given as $S_{RS} \propto P_0 \bar{N}_c$. In the present experiments, to suppress the noise induced by the background light reflected from the vacuum chamber walls and other items in the chamber, a laser beam collimation system and a blackbody-like photo-trap were adopted as shown in figure 18. The dependence of the Rayleigh scattered light signal S_{RS} on the backing pressure P_0, which varies from 0.5 to 1.0 MPa, was described in figure 19. Rayleigh scattered light signal S_{RS} scales with P_0 as $S_{RS} \propto P_0^{1.4}$.

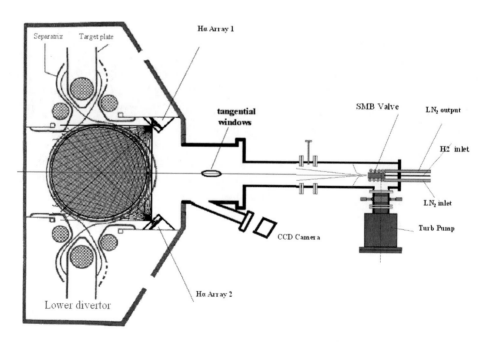

Figure 17. Supersonic molecular beam injection system in the HL-2A Tokamak.

Based on the scaling of $S_{RS} \propto P_0 \overline{N}_c$, the average cluster size therefore varies according to $\overline{N}_c \propto P_0^{0.4}$, deviating essentially from the usual scalings of $\overline{N}_c \propto P_0^{2.0-2.5}$. Normally, a proposed cluster average size $\overline{N}_c \approx 100$ at the "reasonable" onset point of clustering [34], for example, a ratio of signal to noise ≈ 2 may be available, by this view point the onset of clustering for H_2 starts from a much low pressure of 0.1 MPa. In the present experiment $\overline{N}_c \approx 250$ hydrogen atoms at $P_0 = 1.0$ MPa.

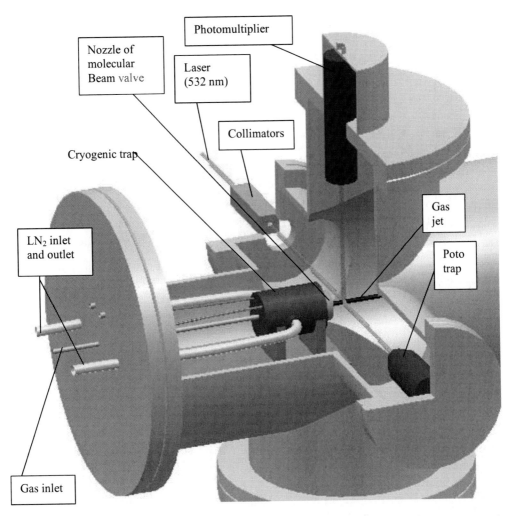

Figure 18. Experimental layout for the size measurement of clusters in a gas jet by a Rayleigh scattering method.

As mentioned in section 3.5, there may be an empirical threshold of the pressure, $P_0 =$ 1.0 MPa, for hydrogen clustering with the series 99 valve in the HL-1M SMBI experiments. The H_α emission signals with some high intensity spikes for multi-pulse SMBI are similar to the case of multi-pellet injection. The characteristic signals may show evidence of the onset of clustering, which resulted in valuable fuelling effects. Hydrogen clusters may occur within the molecular beam under the situation of room temperature and moderate pressure gas, but the cluster size is too small to detect. Due to the limit of the ratio of signal to noise for the Rayleigh scattering method, a cluster with $\overline{N}_c < 100$ atoms can not be detected.

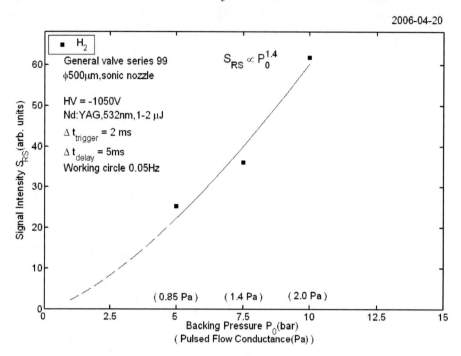

Figure 19. Rayleigh scattered light signal S_R as a function of the gas backing pressure P_0.

4.3   The Fuelling Efficiency of Cryogenically Cooled Gas Cluster Jet

In the HL-2A plasma experiment, the fuelling efficiency for liquid nitrogen temperature gas cluster jet is about 60% under normal wall condition, but it is strongly dependent upon the plasma parameters as well as the particle recycling coefficient. It is in evidence that the cluster jet fuelling effect varies with the gas temperature [35]. For example, under the same gas pressure 1.8 MPa and similar Ohmic discharge parameters, the density increment for shot 4413 with 80 K hydrogen CJI is twice of that for shot 4512 with room temperature gas as shown in figure 20 and 21. In the figures, PuffCtrl(V) is the gas puff control during the discharge, Ph2p is the neutral gas pressure in the main chamber of HL-2A, I_Ha_w is the Ha emission from edge plasma, BOLU08 is heat emission from core plasma and SX10 is soft X-ray emission from core plasma. Even if the gas jet velocity is proportional to square root of gas temperature, the velocity for the low temperature gas cluster is just one half of that for the room temperature one, but the size of cluster produced by cryogenically cooled gas is larger than that produced by room temperature gas. Apparently, the cluster size making a contribution to fuelling effect is more important than the cluster jet velocity. The high fuelling efficiency of the CJI in HL-2A is an evidence, which in accordance with the idea of micro-pellet injection.

4.4   The Injection Depth of the Cluster Jet

The experiment of CJI depth was first carried out on HL-2A, which with divertor. Two diagnostic methods are used to detect the CJI depth, one of them is the tangential $H\alpha$ detection array that shows the trace-ray of CJI location and the other is dTe/dt measured by ECE measurement [36], which shows the electron temperature variation versus the radius due to the injected particle deposition.

Diagnostics method	Shot 4413	Shot 4512
Particles front detected by $H\alpha$ detector array	r = 8 cm	r = 26 cm
Location of maximum dTe/dt	r = 15 cm	r = 21 cm

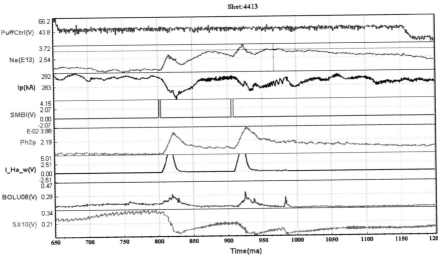

Figure 20. Two pulses of cryogenically cooled gas cluster jet injection into the HL-2A plasma with divertor configuration for shot 4413, the plasma density increases $2.2\times10^{19}m^{-3}$.

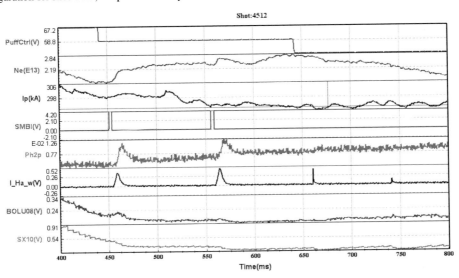

Figure 21. Two pulses of room temperature SMB injection into HL-2A plasma with divertor configuration for shot 4512, the plasma density increases $1.2\times10^{19}m^{-3}$.

The location of maximum dTe/dt means where maximum injected particle number for the jet deposits at, so the distance from the location to the plasma center is far from that from the particles front detected by $H\alpha$ detection array to the center. The injection depth of the low temperature gas CJ is greater than that of the room temperature one under the similar discharge parameters as shown in the following table. It is an other evidence that the condensed hydrogen cluster just likes the micro-pellet, the injection depth is greater than that for the SMBI with same gas pressure but room teperature.

5. INJECTION MODEL OF THE SMB AND CJ

There is a mechanism for interpreting the deeper penetration of the SMB, a simple collective model [37] based on the high flux of neutral particles with a very high density of 4.0×10^{24} m^{-3}, this density is reached below the free jet nozzle when the SMB injector works at 0.5 MPa [38]. When the SMB is injected into a plasma, within which plasma electrons can effectively collide with the dense neutral gas of the surface layer of the SMB, they loose all their energy and finally stop inside this penetration region, so that the neutral gas within this penetration region can protect the bulk particles of the SMB from depositing in the edge plasma zone. So the SMB is surrounded with a cold gas cloud, i.e. a cold protecting channel for SMBI is formed in the injection region. The low temperature region was verified and detected by Langmuir probes and the hollow profiles of electron temperature obtained by ECE measurements during the beam injection give evidence that some of the particles have been injected straight into the center of the plasma. The SMB penetration depths and fuelling profiles have been calculated for HL-1M, JT-60U and ITER-like tokamaks. The calculated value 9 cm of the SMB penetration depth, λ_b, agrees well with the experimental value 8 cm in HL-1M [4]. It should be noted that the number of particles injected is higher at high pressure so the beam density is higher and the shielding effect could be more effective. The cooling passage effect could be enhanced and result in injected particles penetrating into the core region of the plasma, this could in turn lead to a change in the radiation pattern of the recycling hydrogen and the intrinsic impurities like oxygen.

On the other hand, the CJI, in which the average diameter of hydrogen cluster is of nanometer size magnitude, just is a micro-pellet stream injection. A pulse of CJI may be an intensified SMBI and is different from normal PI in series. So the possible explanation for the CJI mechanism is different the PI.

6. SUMMARY

The original reason for proposing SMBI was an attempt to enhance the penetration depth and the fuelling efficiency, as well as to reduce surface adsorption of the injected particles and the impurity content in the plasma [4]. After ten years of practice, SMBI has been developed to become a useful fuelling method that is already considered an improvement over conventional gas puffing and that can be compared with the method of slow or small PI. The high-pressure (more than 1 MPa) multi-pulse SMBI takes a further step towards high performance of the plasma for the onset of injected particle clustering. Normally clustering of the hydrogen jet can be obtained at liquid nitrogen temperature and only a larger cluster (containing more than 100 particles) can be examined by Rayleigh scattering. Hydrogen molecular beam produced from room temperature gas can hardly become large cluster even if at high pressure (say, 8 MPa). Recently the experiment on cryogenically cooled and moderate-pressure hydrogen cluster jet injection into the HL-2A plasma was carried out, and the particle injection depth and the fuelling efficiency were distinctly better than that of the room temperature one.

There are three developing techniques may be a candidate for the fuelling of ITER.

SMBI or CJI, this is the simplest, but effective fuelling method among these techniques. The problem is that the solenoid valve, used to produce the cluster, can not work in strong magnetic field

A pneumatic pulsed SMBI was developed in CEA/DSM/DRFC at Cadarache, which can work in presence of strong magnetic field and shorten the distance between the nozzle of the injector and the edge plasma. The problem is that the pressure of working gas behind the Laval nozzle is not enough high

The opposite way to enhance the injection quality is to increase the fuelling gas temperature and pressure so as to raise the injected particle speed and to vary the injection location from the low magnetic field side to the high field side. This is an attempt for the benefit of the drift effect along the toroidal field gradient just as the new mode of PI in the ASDEX Upgrade tokamak [39]. The problem is that there are a complex pellet producing system and a long bend fly trajectory.

REFERENCES

[1]   Shi, B. R. *Magnetic Confinement Fusion -- Principles And Practice*; Atomic Energy Press: Beijing, CN, 1999; pp 177 (in Chinese).

[2]   Lang, P. T. *et al Phys Rev Lett* 1997, Vol. 79, 1487

[3]   Wagner, F. *et al* Fusion Energy Conference 2002 (*Proc. 19th Int. Conf. Lyon*, 2002) (Vienna: IAEA) CD-ROM file OV/2-4 and http://www.iaea.org/programmes/ripc/physics/fec2002/html/fec2002.htm.

[4]   Yao, L. H. *et al Proc. 20th EPS Conference on Controlled Fusion and Plasma Physics*; Lisbon, 1993, Vol. 17C(I), 303-306.

[5]   Yao, L. H. *et al Nucl Fusion* 1998, Vol. 38, 631-638.

[6]   Yao, L. H. *et al Nucl Fusion* 2001, Vol. 41, 817-826.

[7]   Yao, L. H. *et al Fusion Sci & Tech* 2002, Vol. 42, 107-115.

[8]   Yao, L. H. *et al Nucl Fusion* 2004, Vol. 44, 420-426.

[9]   Gao, X. *et al Nucl Fusion* 2000, Vol. 40, 1875-1883.

[10] Bucalossi, J. et al Fusion Energy Conference 2002 (*Proc. 19th Int Conf. Lyon*, 2002) (Vienna: IAEA) CD-ROM file O-2.07 and http://www.iaea.org/programmes/ripc/physics/fec2002/html/fec2002.htm

[11] Yao, L. H. and Baldzhun J. *Plasma Sci & Tech* 2003, Vol. 5, 1933-1938.

[12] Bucalossi, J. et al *31st EPS Conference on Plasma Physics*, London, 28th June-2nd July 2004, ECA, Vol.28B, P-4, 115-118.

[13] Yan, L. W. *et al Plasma Sci & Tech* 2000, Vol. 2, 431-436.

[14] Scoles, G.. *Atomic and Molecular Beam Methods*; Oxford Univ. Press: New York and Oxford, 1988; Vol. 1, pp15.

[15] Hagena, O. F. *Rev Sci Instrum* 1992, Vol. 63, 2374-2379.

[16] Bernstein, R. B. *Chemical Dynamics via Molecular Beam and Laser Techniques*; Oxford University Press: Oxford, New York, 1982; pp 32-35.

[17] Valyi, L. *Atom and Ion Sources*; Wiley Press: New York, US, 1977; pp 693.

[18] Smith, R. A. Ditmire, T. and Tisch, W. G. *Rev. Sci Instru* 1998, Vol. 69, 3798-3804.

[19] Peng, L. L. *et al Nucl Fusion* 1998, Vol. 38, 1137

[20] Zhang, N. M. J. *et al* J. *Nucl Mater* 1999, Vol. 266-269, 188.

[21] Xiao, Z. G. *et al Fusion Sic & Tec* 2003, Vol. 43, 45-54

[22] Yan, L. W. et al *Nucl Fusion* 2002, Vol. 42, 265-27

[23] Li, K. H. et al *Nucl Fusion Plasma Phys* 1992, Vol. 12, 34-38 (in Chinese)

[24] Bernstein, R. B. *Chemical Dynamics via Molecular Beam and Laser Techniques*; Oxford Univ. Press: Oxford and New York, 1982; pp 32.

[25] Xu, W. B. *Chin Phys Lett* 1998, Vol. 16, 185-188

[26] Dong, J. F. *et al Plasma Phys Control Fusion* 2002, Vol. 44, 371-379.

[27] Yao, L. H. and Baldzuhn, J. *Plasma Sci & Tec* 2003, Vol. 5, 1933-1938.

[28] Hong, W. Y. *et al Chinese Physics Letters* 2002, Vol. 19, 1643-1645.

[29] Field, A. R. *et al Nucl Fusion* 1992, Vol. 32, 1191

[30] Cyrus, S. *et al Rev Sci Instrum* 1992, Vol. 63, 3923.

[31] Klingelhofer, R. and Moser, H. O. *J. Appl Phys* 1972, Vol. 43, 4575.

[32] Liu, Y. *et al Nuc. Fusin* 2005, Vol. 45, S239.

[33] Lei, A.L. *et al Chin Phys Lett* 2000, Vol. 17, 661.

[34] Buch, U. and Krohne, R. *J. Chem Phys* 1996, Vol.105, 5408.

[35] Yao, L. H. *et al* Fusion Energy Conference 2006 (*Proc. 21th Int. Conf. Chengdu*, 2006) (Vienna: IAEA) and http://www.iaea.org/programmes/ripc/physics/fec2006/html/ fec2006.htm.

[36] Sun, H. J. *et al 33st EPS Conference on Plasma Physics*, Roma, 19th -23nd June, 2004.

[37] Song, X. M. *et al J. Plasma Fusion Res* 2000, Vol. 76, 282-287.

[38] Ditmire, T. *et al Optics Letters* 1998, Vol. 23, 618

[39] Lang, P. T. *et al Nucl Fusion* 2001, Vol. 41, 1107

In: Advances in Plasma Physics Research, Volume 7   ISBN: 978-1-61122-983-7
Editor: Francois Gerard   © 2011 Nova Science Publishers, Inc.

Chapter 8

A Diagnostic Method of Electromagnetic Field Patterns of Fast Wave in High Temperature Plasma

Mikio Saigusa[*] and Sadayoshi Kanazawa[1]

Ibaraki University, 4-12-1, Nakanarusawa, Hitachi-shi, Ibaraki-ken, Japan,
[1]Present affiliation: Panasonic AVC Networks Company,
Matsushita Electric Industrial Co., Ltd., Osaka-fu, Japan

Abstract

Physics of fast wave current drive and electron heating at a frequency range of ion cyclotron resonance have been investigated in fusion devices. The theoretical studies predicted the fine structure of electromagnetic field of fast waves in fusion plasmas. However, behaviors of fast waves in plasmas can be estimated from macroscopic experimental results, for example, the reduction of toroidal loop voltage and the motional stark effect with the poloidal magnetic field. In this chapter, the new method for diagnosing electromagnetic field pattern of fast waves in toroidal plasmas is introduced. The fluctuation of ponderomotive potential at a beat wave frequency between the two frequency fast waves produces the actual potential fluctuation via electron and ion motions in fusion plasmas. The amplitude of potential fluctuations are proportional to the square of electric field strength of fast waves. Therefore, the amplitude pattern of potential fluctuations at a beat wave frequency indicates the field pattern of fast wave in plasmas. The potential fluctuation patterns in high temperature plasma can be detected with a heavy ion beam probe (HIBP), so that the pattern of electromagnetic field of fast wave can be estimated from the data of HIBP with the beat wave technique. The potential fluctuation pattern at the beat wave frequency (90 kHz) has been measured with HIBP system during the fast wave pulses at a frequency of 200 MHz in JFT-2M, which is a middle size non-circular tokamak. The measured levels of potential fluctuations decrease with increasing an electron temperature, consistent with the improvement of wave absorption. The measured potential fluctuation levels are similar to the theoretical

---

[*] E-mail address: saigusa@ee.ibaraki.ac.jp

predictions. The feasibility of this method in large fusion devices depends on the wave absorption and the sensitivity of HIBP.

1. INTRODUCTION

Physics and techniques of fast wave current drive (FWCD) and direct electron heating at a frequency range of high harmonic ion cyclotron resonance have been investigated in tokamaks [1] and spherical tokamaks [2]. Many excellent theoretical physicists predict the fine structure of electromagnetic field of fast waves in toroidal plasmas [3]. However, the detail information of propagation and absorption of fast waves in plasmas can be estimated from macroscopic experimental results, for example, the reduction of toroidal loop voltage and the motional stark effect with the poloidal magnetic field, together with complicated calculations using many time slices for evaluating the internal structure of electric field induced by fast wave driven current [4].

On the other hand, the wave number of the fast wave has been measured with a phase contrast imaging method via the electron density fluctuations in Alcator C-Mod, which proved the propagation of fast wave into core plasma [5]. However, the spatial electromagnetic field pattern of the fast wave has never been diagnosed in a toroidal plasma. The measurement of an accurate electromagnetic field pattern is useful for comparing with theoretical predictions. In this chapter, the new diagnostic method for estimating electromagnetic field pattern of fast waves in toroidal plasmas is introduced, which uses the ponderomotive potential fluctuations of beat waves [6]. The various application of radio frequency waves in a fusion plasma was introduced in section 2. The general theory of ponderomotive force is explained in section 3. The basic principle of the diagnostics is explained in section 4. The feasibility of diagnostics was proved in JFT-2M, which was a mid sized non-circular tokamak with a FWCD system, ECH system, and a HIBP system. These experimental results are showed in section 5. The comparison of theoretical predictions and experimental results are shown in section 6. The future prospect of this diagnostics in large tokamaks are discussed in section 7. Finally, all topics are summarized in section 8.

2. APPLICATION OF RADIO FREQUENCY WAVES IN A FUSION PLASMA

Up to now, many plasma waves have been adopted for additional plasma heating and driving non-inductive current in fusion plasmas. The electron cyclotron (EC) pre-ionization is the one of most popular methods to ionize gas for plasma formations. The electron cyclotron resonance heating (ECH) and current drive (ECCD) is one of the most promising heating and non-inductive current driving method, because of local heating and controllable current profile, and the wide clearance between antenna and plasma edge. On the other hand, the ion cyclotron resonance heating (ICH) is the easiest and cheapest way to heat ions and electrons in large fusion devices, which demands the machine size for avoiding ion orbit loss. Lower hybrid (LH) wave is the most suitable wave for driving non-inductive current in a toroidal plasma, as if the electron temperature is too low to drive non-inductive current for the other methods. Ion Bernstein waves is useful for ion and electron heating and improving the particle transport in tokamak plasma [7, 8]. Electron Bernstein waves are useful for inside launch electron cyclotron heating via mode conversion from an extraordinary wave around the upper hybrid resonance layer. In addition to that, the electron Bernstein waves are started to be adopted for electron heating in spherical tokamaks, because the magnetic field is too low for the propagation of ordinary and extraordinary waves [9].

In 2001, EC and IC systems are planed for pre-ionization, electron and ion heating and driving current at the first phase of ITER project , and the LH system is planned in the second phase as shown in Table 1 [10]. Besides the rf systems, the negative neutral beam system (beam energy of 1 MeV, beam power of 33MW) has been planed for heating and current drive in ITER-FEAT. The wave dumping of fast waves is expected to be complicated due to the existence of various impurity ions, which have many high harmonic resonance layers in plasma.   Therefore, it is important for high efficiency IC heating to diagnose of the wave field patterns in ITER-FEAT plasma, especially.

Table 1. heating and current drive radio frequency systems for ITER-FEAT design

	Ion Cyclotron Heating &CD System (40-55MHz)	Electron Cyclotron Heating &CD System (170GHz)	Electron Cyclotron Pre-ionization System (120GHz)	Lower Hybrid Current Drive System (5GHz)
Power injected per unit equatorial port (MW)	20	20	2	20
Number of units for the first phase	1	1	1	0
Total Power for the first phase (MW)	20	20	2	0

In addition to that, the conditions of rf-plasma coupling, propagation and dumping of plasma waves depend on the various plasma parameters and wall conditions in each frequency range. Therefore, the experimental physicist often can't find what is wrong in actual experiments, as if the reaction of heating or current drive by rf waves have never been observed in macroscopic plasma parameters for several months. However the diagnostic method for drawing electromagnetic field pattern have never been invented by the end of the twenty century. We would like to propose the new diagnostics for obtaining wave field patterns of high harmonic fast wave in fusion plasmas in order to understand the wave physics in plasma. This study is the first step for following developments of excellent diagnostics in order to obtain accurate field pattern of various plasma waves in reactor grade fusion plasma.

3. PONDEROMOTIVE FORCE AND POTENTIAL

Ponderomotive force is the most popular non-linear force in plasma physics, which is a kind of radiation pressure of the electromagnetic wave. The pressure of light is usually negligible small in vacuum. On the other hand, it has same effect in plasma via mainly electron motions.

The typical example is the self focusing phenomena of intensive laser beam in plasma. The ponderomotive force of electromagnetic field of laser beam can push out the plasma from its channel, and the hollowed profile of refractive index focuses the laser beam by itself. The recent advanced concept for inertia fusion project: "First Ignition" which demands self focusing phenomena of intensive laser in relativistic plasma was proposed by the researchers in Osaka university.

In the case of non-magnetized plasma, the physical picture and simple formula have been explained clearly in Ref [11]. The equation of motion for a charged particle in electromagnetic field is

$$mdv/dt = q[E(r)+v \square B(r)] \qquad (1),$$

where q, $\square v \square \square \square B$ are an electric charge, electric field, a velocity of charged particle, and magnetic field, respectively.

If the electromagnetic field is assumed to be $E=E_{rf}(r)cos\square t$, the first order term of the equation of motion is given as $mdv_1/dt = qE(r)$, so that first order velocity:

$$v_1 = -(q/m\square)E_{rf}sin\square t = dr_1/dt \text{ and } \square r_1 = (q/m\square^\square)E_{rf}cos\square t.$$

The B_{rf} can be obtained to be $B_{rf} = -(1/\omega)\nabla \times E_{rf}|_{r=r_0}$ $sin\omega t$ from the equation of $\nabla \times E = -dB/dt$.

The ponderomotive force originates in the second order term of the equation of motion for a charged particle in electromagnetic field ($mdv_2/dt = q((\square r_1 \cdot \square_. \square v_1 \square B_{rf}))$, where $\square r_1$ B_{rf} are a displacement of charged particle and rf magnetic field, respectively. Averaging over time, we could have the equation for a single charged particle :

$$F = -\frac{q^2}{2m\omega^2}\left[\left(E_{rf} \cdot \nabla\right)E_{rf} + E_{rf} \times \left(\nabla \times E_{rf}\right)\right]$$

$$= -\frac{1}{4}\frac{q^2}{m\omega^2}\nabla|E_{rf}|^2 \qquad (2).$$

The time averaged ponderomotive force can be understood as the pressure of electromagnetic wave. In quantum mechanics, it is the momentum: $\hbar k$ which is in the same direction of the wave vector: k. From another point of view, the charged particle oscillate in the electric field direction, while the rf magnetic field distorts their orbits. Thus, the particles are pushed by Lorentz force: $qv \times B_{rf}$ to the direction of k.

The ponderomotive force at the beat wave frequency between two frequency waves working on the single charged particle can be written as

$$F_{beat} = -\frac{1}{2}\frac{q_j^2}{m_j\omega_1\omega_2}\nabla(E_1 \cdot E_2^*) \qquad (3),$$

where ω_1, ω_2 are angular frequencies of two waves, q_j and m_j are an electric charge and a mass of j particle, the marks "*" indicate the complex conjugate.

In a magnetized plasma, the ponderomotive force was studied by Motz and Watson in Ref. [10]. Since the effect of guiding center theory modified the ponderomotive force, the "rf quasi-potential" of the electron is changed from $\psi = \dfrac{e^2}{m\omega^2}|E|^2$ to

$$\psi = \frac{e^2}{m\omega}\left[\frac{|E_L|^2}{\omega + \omega_{ce}} + \frac{|E_R|^2}{\omega - \omega_{ce}} + \frac{|E_z|^2}{\omega}\right] \qquad (4),$$

where a magnetic field line is parallel to the z axis, and $E_L \equiv (E_x + iE_y)/\sqrt{2}$, $E_R \equiv (E_x - iE_y)/\sqrt{2}$.

The ponderomotive force on an electron is proportional to $-\nabla\psi$. The effect of the resonant denominators : $1/(\omega - \omega_{ce})$ may enhance the amplitudes of rf electric field, when ω is near a electron cyclotron angular frequency(ω_{ce}), but not so close that $(\omega - \omega_{ce})/\omega \approx v/c$.

4. BASIC PRINCIPLE OF DIAGNOSTICS FOR ELECTROMAGNETIC FIELD OF FAST WAVE IN FUSION PLASMA

The fluctuation of the ponderomotive force produced by the beat wave generated from two waves propagates at a group velocity of $(\omega_1 - \omega_2)/(k_1 - k_2)$, where the angular frequencies of ω_1 ω_2 and the wave vectors of k_1 and k_2, correspond to the two waves. This propagating beat wave can generate a force through "nonlinear Landau damping", which is the wave- particle resonance phenomenon at the group velocity: $(\omega_1 - \omega_2)/(k_1 - k_2)$.

The principle of the diagnostic method for measuring the spatial pattern of the fast wave electromagnetic field consists of three steps. First, the beat wave between the two fast waves (ω_1 and ω_2)produces a fluctuation of the ponderomotive potential which is different work on electrons and ions [11,12,13]. Next, potential fluctuations are produced by the electron and ion motions. Finally, the spatial structure of the potential fluctuations patterns can be measured at the beat wave frequency with a heavy ion beam probe (HIBP) [14]. When the beat wave frequency ($\omega_1 - \omega_2$) is much less than the center frequency ($(\omega_1 + \omega_2)/2$), the pattern of the fluctuation amplitude of the ponderomotive forces at a beat wave frequency are thought to be the same pattern of the square of the fast wave electric field, when both of the fast waves are radiated from the same antenna at almost the same power level.

When the two fast wave frequencies are almost the same, the ponderomotive force on the charged particles produced by the beat wave can be written as

$$= -(q_j^2/2m_j)\nabla[E_{R1}\cdot E_{R2}*/((\omega+\varepsilon_j\omega_{cj})\cdot\omega)+E_{L1}\cdot E_{L2}*/((\omega-\varepsilon_j\omega_{cj})\cdot\omega)$$
$$+E_{z1}\cdot E_{z2}*/\omega^2]\cdot\exp[i\{(k_1-k_2)\cdot r-(\omega_1-\omega_2)t\}]$$

(5),

where ω is fast wave angular frequency ($\omega \sim \omega_1$, ω_2), ω_{cj}, q_j and ε_j are a cyclotron angular frequency, an electric charge and a sign of electric charge of j particle, the marks "*" indicate the complex conjugate, respectively [11, 13]. E_{R1} (or E_{R2}) and E_{L1} (or E_{L2}) are the perpendicular electric field components of right and left handed circularly polarized waves, respectively, and E_{z1} (or E_{Z2}) is the parallel electric field component to the magnetic field line, where the static magnetic field is taken to be in the z-direction.

The ponderomotive potential: ϕ_{pj} of beat wave can be defined by the followings:

$$F_j = q_jE = q_j(-\nabla\phi_{pj})$$

(6),

Therefore

$$\phi_{pj} = (q_j/2m_j)[E_{R1}\cdot E_{R2}*/((\omega+\varepsilon_j\omega_{cj})\cdot\omega)+E_{L1}\cdot E_{L2}*/((\omega-\varepsilon_j\omega_{cj})\cdot$$
$$+E_{z1}\cdot E_{z2}*/\omega^2]\cdot\exp[i\{(k_1-k_2)\cdot r-(\omega_1-\omega_2)t\}]$$

(7).

The ponderomotive potential is proportional to the ratio of electric charge to mass of particle, so that it mainly works on electrons, except nearby ion cyclotron resonance. Note, the heavy ions for diagnostic beam see a small ponderomotive force, since that the ratio of electric charge to mass is small.

The high harmonic fast wave is a right handed elliptically polarized wave in an oblique propagation. However, the right handed elliptically polarized wave still consists of E_{Rj} and E_{Lj} components. Therefore, $(E_R/E_Z)^2$ and $(E_L/E_Z)^2$ in cold plasma theory can be shown as the followings:

$$\frac{E_R^2}{E_z^2} = \frac{(N^2-L)^2(N^2-N_{//}^2-P)^2}{2(N^2-S)^2N_{//}^2(N^2-N_{//}^2)}$$

(8),

$$\frac{E_L^2}{E_z^2} = \frac{(N^2-R)^2(N^2-N_{//}^2-P)^2}{2(N^2-S)^2N_{//}^2(N^2-N_{//}^2)}$$

(9),

where $N^2 = (-b \pm\sqrt{b^2-4c})/2$, $b=(P/S-1)\cdot N_{//}^2 - (P+ RL/S)$, and $c=(RL/S - P)\cdot N_{//}^2 + PRL/S$. P, S, D, R, and L are the popular Stix's notation as the followings,

$$P = 1 - \sum_j \frac{\omega_{pj}^2}{\omega^2}, \quad S = 1 - \sum_j \frac{\omega_{pj}^2}{\omega^2 - \omega_{cj}^2}, \quad D = \sum_j \frac{\varepsilon_j \omega_{cj} \omega_{pj}^2}{\omega(\omega^2 - \omega_{cj}^2)}, \quad R = S + D, \text{ and } L = S - D,$$

where ω_{pj} is a plasma frequency. Then, the ratio of the ponderomotive potential on ions (ϕ_{pi}) to that on electrons (ϕ_{pe}) can be estimated to be,

$$\frac{\phi_{pi}}{\phi_{pe}} = \frac{q_i}{-e} \cdot \frac{m_e}{m_i} \cdot \frac{\dfrac{E_R^2}{E_Z^2} \cdot \dfrac{\omega}{(\omega + \omega_{ci})} + \dfrac{E_L^2}{E_Z^2} \cdot \dfrac{\omega}{(\omega - \omega_{ci})} + 1}{\dfrac{E_R^2}{E_Z^2} \cdot \dfrac{\omega}{(\omega - \omega_{ce})} + \dfrac{E_L^2}{E_Z^2} \cdot \dfrac{\omega}{(\omega + \omega_{ce})} + 1}$$

(10),

where the sign of ϕ_{pi} and ϕ_{pe} indicate the phase information of the fluctuation.

Figure 1 shows ϕ_{pi}/ϕ_{pe} as a function of frequency at typical plasma parameters (B=1 [T], electron density =1.5 × 10^{19} [m^{-3}], deuterium plasma) for the JFT-2M FWCD experiments. ϕ_{pi} has a singular point at $\omega = \omega_{ci}$, that is 7.6 MHz, so that the Eq. (10) has is not correct around the resonance frequency (below 10 MHz). The ϕ_{pi}/ϕ_{pe} decreases with increasing the frequency, and it becomes 0.12 at a frequency of 200 MHz (~26ω_{ci}). Figure 2 indicates the effect of each electric field component (mainly E_R and E_L on ϕ_{pi} and ϕ_{pe}, where all terms are normalized by ϕ_{pe} from E_Z. ϕ_{pi} from E_Z is about - 1/3670 in this frequency range, while the other terms are much larger than unity and they are the functions of a frequency. ϕ_{pe} from E_R and ϕ_{pe} from E_L have opposite signs to each other, and ϕ_{pe} from E_R is larger than ϕ_{pe} from E_L. The signs of ϕ_{pi} from E_R and ϕ_{pi} from E_L are the same of the sign of ϕ_{pe} from E_R. Therefore, the ponderomotive potential from the beat waves applies mainly to the electrons and is proportional to the square of the perpendicular electric field components at the frequency range of 200 MHz. The electric potential fluctuation at the beat wave frequency is then caused by the perpendicular electron motions affected by the ponderomotive force.

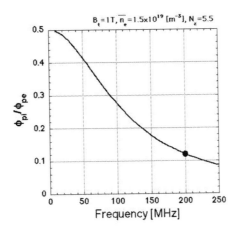

Fig. 1 ϕ_{pi}/ϕ_{pe} produced by beat wave against the frequency at the typical plasma parameters in JFT-2M FWCD experiments, where a magnetic field is 1 [T], an electron density is 1.5 × 10^{19} [m^{-3}], the ion species is pure deuteron. ϕ_{pi}/ϕ_{pe} is about 0.12 at the frequency of 200 MHz, which is the frequency of FWCD system in JFT-2M (~26 ω_{ci}).

Full wave theory can predict the fine structure of the electromagnetic field of high harmonic fast waves. The poloidal electric field patterns of the fast wave in a poloidal cross section predicted with the full wave code (TASK/WM) considering 64 poloidal modes are shown in Figs. 3(a), 3(b) and 3(c) for the central electron temperature :T_{e0} =6 keV, 3 keV, and 1.5 keV, respectively [15]. The toroidal refractive index of excited power spectrum on plasma outer surface is assumed to be 5. The target plasmas are deuterium plasmas, a toroidal magnetic field on magnetic axis of 1.15 [T], and the electron density and the electron temperature profiles of $n_e(r)=(1-(r/a)^2)^{0.5}+1$ [10^{19}m^{-3}], $T_e(r)$ =$T_i(r)$=$(T_{e0}-T_{eb})$ $(1-(r/a)^2)$ + T_{eb}. The fast wave radiated from the antenna installed on the outer lower wall is propagating into a core plasma as a traveling wave with strong electron Landau damping for T_{e0} of 6 keV, while a weak standing wave is seen for T_{e0} of 3 keV, and a strong standing wave structure is seen for T_{e0} of 1.5 keV. The predicted peak poloidal electric field components at the radiated fast wave power of 200 kW for T_{e0} of 6 keV, 3 keV and 1.5 keV are 3.8 kV/m, 12.3 kV/m, and 130 kV/m, respectively. These results suggest the possibility that the structure of ponderomotive potential produced by the strong electric fields can be diagnosed in lower electron temperature plasma, even for low power FWCD operations in JFT-2M.

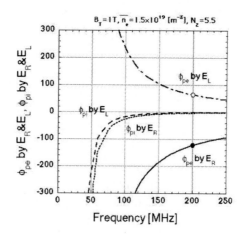

Fig. 2 ϕ_{pi} with E_L, ϕ_{pi} with E_R, ϕ_{pe} with E_L, and ϕ_{pe} with E_R are plotted against a frequency at the same plasma parameters of Fig. 1, where they are normalized by the ϕ_{pe} with E_Z. The ϕ_{pi} with E_Z can be neglected.

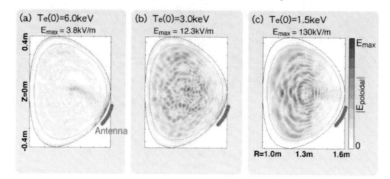

Fig. 3 The electric field patterns of the fast wave on the poloidal cross sections predicted with the full wave code calculations for (a) T_{e0}=6 keV, (b) T_{e0}=3 keV, and (c) T_{e0}=1.5 keV. Excited toroidal refractive index (N_z) on plasma outer surface is assumed to be 5. Main plasma parameters are $B_T(0)$=1.15 [T], a central electron density =2.0_×_10^{19} [m^{-3}], $T_e(r)$ =$T_i(r)$=$(T_{e0}-T_{eb})$ $(1-(r/a)^2)$ + T_{eb}, deuterium ion plasma.

5. EXPERIMENTAL RESULTS IN JFT-2M

For the comparison of the theoretical predictions and the experimental results, the electromagnetic field patterns of fast wave are measured for various plasma parameters in JFT-2M. The target deuterium plasma parameters are lower single null divertor configurations, $R/a \sim 1.32$ m/0.30 m, $Ip = 130$ kA for co-FWCD and $Ip = 140$ kA for counter FWCD, $B_t(0) \sim 1.15$ T, line averaged electron density $1.8 \sim 2.3 \times 10^{19}$ m^{-3}, and the coupled fast wave power of about 200 kW. A schematic of the FWCD system for these experiments is shown in Fig. 4. The two frequencies are (199.91 MHz and 200.00 MHz) generated by two oscillators and combined with a hybrid junction. The signal is then amplified, transmitted and radiated from a combline antenna as shown in Fig. 5 [16]. The combline antenna is a kind of traveling wave antennas, which was proposed by Moeller in General Atomics. The features of combline antenna are an excitation of sharp toroidal refractive index spectrum and good impedance matching to plasma. The combline antenna with 12 current straps in JFT-2M has a frequency band width of about 10 MHz with a dummy antenna load at a central frequency of 200 MHz [17]. The peak toroidal refractive index is estimated to be $N_z \approx 5$ with $\Delta N_z \approx 1.6$ and ~80 % of the total power radiated in the desired direction from the coupling code. Additional electron heating was supplied by an ECH system consisting of three gyrotrons, each gyrotron generating about 200 kW at a frequency of 60 GHz. Three gyrotrons were used for these experiments. The electron cyclotron resonance layer is located around r/a~0.3 in low field side as shown in Fig. 6. A projection of the heavy ion beam orbits (primary beams and secondary beams) on the poloidal cross section is shown in Fig. 6. The HIBP system in JFT-2M using singly ionized thallium ion (Tl$^+$) beam at an acceleration voltage of 350 keV can measure the potential fluctuations and the density fluctuations with a spatial resolution of about 20 mm [18]. Figure 7 indicates the basic principle of the HIBP. The secondary beams of HIBP have the local information of the potential fluctuation at the charged point due to the difference of the Larmor radius: q_iB/m_i between Tl$^+$ and Tl^{2+}. The highest sampling rate and the record window length are 1M samples/s and 128 kpoints, respectively. The heavy ion beam can be scanned in the poloidal direction during a fast wave pulse.

A frequency spectra of the signal from an rf probe installed in an adjacent port from the antenna port during a fast wave pulse are shown in Figs. 8(a) for vacuum and 8(b) for plasma. These signals indicate the peak values during the rf pulse. The two frequency signals are clearly observed at the excited frequencies of 200.0 MHz and 199.91 MHz in both conditions. The other signal levels are - 40 dB lower than the main two signals. Figure 9 shows the frequency spectrum of a potential fluctuation measured with the HIBP system during a fast wave pulse at the fixed radial position (r/a~0.36). The sampling period for the fast Fourier transform (FFT) is 32 ms at a sampling frequency of 500 kHz. The beat wave between the two fast waves is clearly observed at the frequency of 90 kHz. The signal to noise ratio is 10.

The time evolutions of a potential signal with HIBP at a radial position of r/a~0.36, a frequency spectrum of its potential signals, a central electron temperature measured with soft X-ray energy spectra and the line averaged electron density are shown in Figs. 10(a), 10(b) and 10(c). The white lines in Fig. 10(b) indicate the pulse durations of fast wave (from 635 to 835 ms) and ECH (from 750 to 835 ms). The frequency spectrum of 90 kHz, which is the beat wave frequency, can be clearly observed during fast wave pulse duration at the coupled power level of about 200 kW. The line averaged electron density increases after the onset of fast wave pulse and ECH pulse (~330 kW). The central electron temperature increases from about 0.8 keV to about 1.4 keV after the onset of ECH pulse, and decreases gradually during ECH pulses, probably due to the increasing electron density cutting off electron cyclotron heating (the cutoff density for the second harmonic extraordinary wave is n_e=2.2 $\square$ 10^{19} m^{-3}).

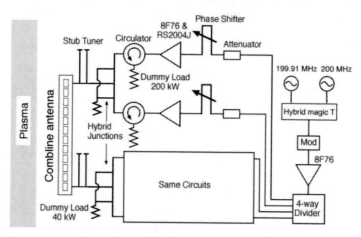

Fig. 4 A schematic diagram of FWCD system for these experiments. Either of co- or counter- traveling wave can be excited easily by controlling the attenuators on the transmission lines.

Fig. 5 A combline antenna attached on vacuum vessel in JFT-2M.

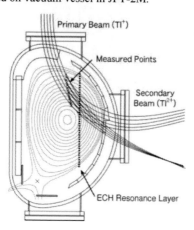

Fig. 6 Schematic view of Tl$^+$ ion beam trajectories of the HIBP on a poloidal cross section in JFT-2M plasma. Secondary beams have the local information of the potential fluctuations at the measured points shown as the cross points. The dotted line indicates the electron cyclotron resonance layer.

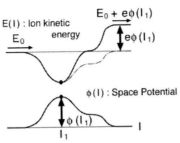

Fig. 7 Basic principle of the HIBP. The difference of ion kinetic energy between before and after crossing a plasma: $e\phi(I_1)$ indicates the local space potential at the ionized point.

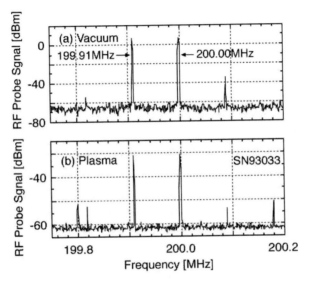

Fig. 8 Frequency spectrum of rf probe signals (a) in vacuum and (b) in a plasma. The expected rf signals radiated from the combline antenna are observed at the frequencies of 200.0 MHz and 199.91 MHz in both conditions.

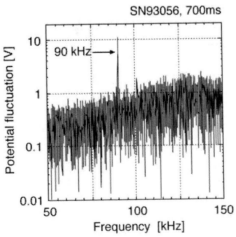

Fig. 9 Frequency spectrum of the potential fluctuation measured with the HIBP system during the fast wave pulse at the fixed radial position (r/a~0.36). The signal at the frequency of 90 MHz indicates the potential fluctuation produced by the fluctuation of ponderomotive potential at a beat wave frequency.

6. COMPARISON OF EXPERIMENTAL RESULTS AND THEORETICAL PREDICTIONS BY FULL WAVE CODE

The pattern of poloidal electric field components of the fast wave at a frequency of 200 MHz in a poloidal cross section are predicted with the full wave code (TASK/WM) as shown in Fig. 11, where the full wave code calculated three components of electric field. The used plasma and rf parameters of JFT-2M and FWCD system are shown in Table 2. The used profiles of electron temperature and electron density are assumed to be $T_e(r)=(T_{e0}-T_{eb})(1-(r/a)^2)+T_{eb}$ [keV] and $n_e(r)=(1-(r/a)^2)^{0.5}+1$ [10^{19} m^{-3}], where the high electron density in the edge region is based on the experimental data. The ponderomotive potentials are estimated to be $\phi = \phi_{pi} + \phi_{pe}$ from the Eq. (7), neglecting the effect of impurity ions and the damping of the potential fluctuation. The cross points indicate the measured points with the HIBP system. The calculation points from the full wave code used for the comparison between the theory and the experimental results are located along the solid line shown in Fig. 11. The strong eigen mode structure can be seen in the radial and poloidal directions, because of the weak electron Landau damping due to low electron temperature.

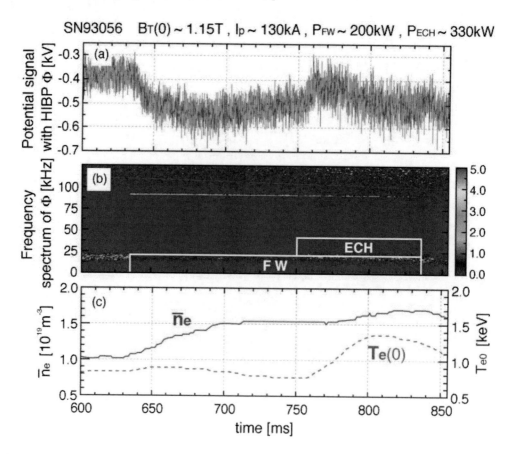

Fig. 10 Time evolutions of (a) a potential signal with HIBP system at the fixed radial position (r/a~0.36), (b) the frequency spectrum of the above potential signal, (c) the central electron temperature estimated from the soft X-ray energy spectrum and the line averaged electron density.

Table 2. The used plasma and rf parameters of JFT-2M for the full wave code (TASK/WM) calculations

Main Plasma Parameters	
Plasma Gas	D_2
Major Radius: R	1.29 m
Minor Radius: a	0.296 m
Vertical Elongation :b/a	1.29
Triangularity: δ	0.289
Central Safety Factor: $q(0)$	1.0
Surface Safety Factor: $q(a)$	4.982
Plasma Current: I_P	130 kA
Toroidal Magnetic Field: $B_T(0)$	1.161 T
Central Electron Density: n_{e0}	2.0×10^{19} m^{-3}
Central Electron Temperature: T_{e0}	1.5 keV
Edge Electron Temperature: T_{eb}	0.08 keV
RF Parameters	
Frequency :f	200 MHz
RF Power: P_{RF}	200 kW
Toroidal Refractive Index: N_z	5

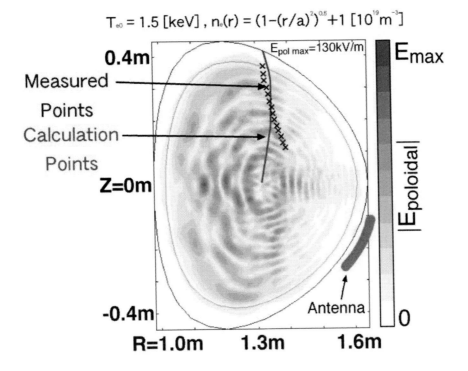

Fig. 11 The poloidal electric field pattern of the fast wave at a frequency of 200 MHz in a poloidal cross section predicted with full wave code (TASK/WM), where T_{e0}=1.5 keV and $n_e(r) = (1-(r/a)^2)^{0.5}+1$ $[10^{19}$m$^{-3}]$. The cross points indicate the measured points with the HIBP system and the solid line indicates the points of numerical calculations for comparing with the experimental data as shown in Fig. 10.

The radial profiles of the amplitude of potential fluctuation at a beat wave frequency are shown in Fig. 12. The closed triangles and the solid line show the prediction with the full wave code and a fitting curve at the same parameters in Fig. 11. The open circles and the dashed line show the measured amplitudes and the fitting curve during fast wave pulses (at $t\sim0.7$ s, $T_e(0)\sim0.8$ keV). The close circles and the dotted line show the measured amplitudes and the fitting curve during the combined operation of fast wave and ECH pulses (at $t\sim0.8$ s, $T_e(0)\sim1.4$ keV). The noise level on the measured signal is about 1V. The measured amplitude of the potential fluctuations for $T_e(0)\square0.8$ keV are clearly larger than those for $T_e(0)\square1.4$ keV at all measured radial points ($r/a=0.36\sim0.94$). In both conditions of co- and counter-FWCD experiments, the tendency for the fluctuation amplitudes to decrease with increasing the electron temperature was seen. These results suggest that the amplitudes of the potential fluctuations decrease with the improvement of the wave absorption caused by the increase in electron Landau damping. The levels of the measured amplitude of potential fluctuations are similar to the theoretical predictions. However, the measured data are flat in radial direction, while the theoretical prediction indicates the strong standing wave structure in a radial direction, with a period (a half wavelength) of about 2 cm. This contradiction might be due to the HIBP spatial resolution of about 20 mm and a neglect of a N_z spectrum width in the full wave code.

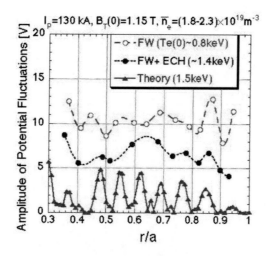

Fig. 12 The radial profile of the fluctuation amplitude of ponderomotive potential at the beat wave frequency. The closed triangles and the solid line show the predictions with the full wave code calculations and the fitting curve. The open circles and the dashed line show the measured amplitude points and the fitting profile during a fast wave pulse. The close circles and the dotted line show the measured amplitude points and the fitting profile during the combined operation of fast wave and ECH ($2\omega_{ce}$), where the electron cyclotron resonance layer is located on $r/a\sim0.3$.

7. FOR LARGE TOKAMAKS AND ALTERNATIVES

It have been proved that the potential fluctuation pattern at the beat wave frequency of the fast waves have been measured with HIBP system in FWCD experiments in middle size tokamaks. However, this method demands intensive electric field due to the high rf power or weak wave dumping. For confirming the feasibility of this diagnostic method for large tokamaks, the potential fluctuation level of the beat wave between two fast waves is evaluated using the typical plasma parameters of JT-60U, which is the large tokamak in Japan. Table 3 shows the used plasma and rf parameters of JT-60U for the calculations with the full wave code (TASK/WM).

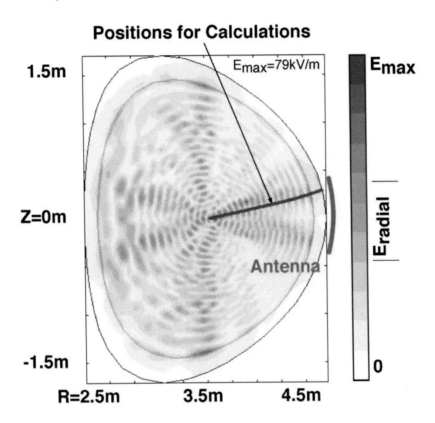

Fig. 13 The radial electric field pattern of the fast wave at a frequency of 112 MHz, the toroidal refractive index (N_z) of 2 and a rf power of 20MW in a poloidal cross section of JT-60U predicted with full wave code (TASK/WM), where the used plasma parameters are assumed to be $B_T(0)$=3.8 [T], $T_e(0)$=8.0 [keV] and $n_e(0)$ =3.0 [10^{19}m^{-3}]. The adopted heating regime is the second harmonic minority heating at the minority (H) concentration of 3% in D plasma. The solid line indicates the points of numerical calculations for comparing with the experimental data as shown in Fig. 12.

Figure 13 shows the patterns of radial components of electric field of the fast wave on the poloidal cross section of JT-60U, where the profiles of electron temperature and electron density are assumed to be $T_e(r)=(T_{e0}-T_{eb})(1-(r/a)^2)+T_{eb}$ [keV] and $n_e(r) = (2.7-(r/a)^2)^{0.5}+0.3$ [10^{19}m^{-3}], respectively. The adopted heating regime is the second harmonic minority heating at the minority ions (H) concentration of 3% in the majority ions(D). The feature of this heating regime is energetic ion tail heating which is suitable for the experimental simulation of alpha particle heating in fusion reactor. The solid line indicates the points of numerical calculations for comparing with the experimental data. The intensive eigen mode structure can be observed in Fig. 13, because the wave absorption is weak due to low minority concentration for second harmonic minority heating regime.

Table 3. The used plasma and rf parameters of JT-60U for full wave code (TASK/WM)

Main Plasma Parameters	
Major Radius: R	3.5 m
Minor Radius: a	1.10 m
Vertical Elongation :b/a	1.32
Triangularity: δ	0.28
Central Safety Factor: $q(0)$	1.49
Surface Safety Factor: $q(a)$	6.75
Plasma Current: I_P	1.7 MA
Toroidal Magnetic Field: $B_T(0)$	3.8 T
Central Electron Density: n_{e0}	3×10^{19} m^{-3}
Central Electron Temperature: T_{e0}	8.0 keV
Edge Electron Temperature: T_{eb}	0.16 keV
RF Parameters	
Frequency :f	112 MHz
RF Power: P_{RF}	20 MW
Toroidal Refractive Index: N_z	2

Figure 14 indicates the expected radial profile of the fluctuation amplitude of ponderomotive potential at the beat wave frequency in JT-60U. The closed circles shows the predictions with the full wave code calculations, where the solid line is the fitting line with cubic spline functions. The amplitude of potential fluctuation in central region is higher than that in peripheral region, not due to near the second harmonic cyclotron resonance, but due to the eigen mode structure as shown in Fig. 13. The expected levels of the amplitude of the potential fluctuation are lower than those for JFT-2M experiments, probably due to the stronger wave dumping. The stronger wave dumping in JT-60U is thought to be caused by the higher electron temperature and the lower harmonic number of the ion cyclotron resonance. Therefore, the electric field of fast wave will be able to diagnosed in large fusion devices only for weak wave dumping or high sensitivity of HIBP.

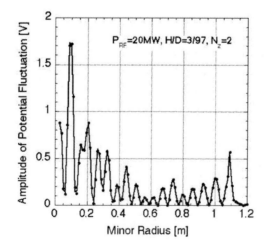

Fig. 14 The radial profile of the fluctuation amplitude of ponderomotive potential at the beat wave frequency. The closed circles show the predictions with the full wave code calculations. The minority ion (H) concentrations in the majority ions (D) are 3%.

The beat wave technique for diagnostic rf field pattern can be extended to other rf heating and current driving methods using high power amplifiers. The diagnose of rf field pattern of LHCD and IBW heating is applicable, because the high power amplifiers of LHCD system are usually klystron amplifiers, and the high power amplifiers of IBW system are tetrode amplifiers. On the other hand, EC system demands a high power gyrotron which is a kind of oscillator, so that the beat wave frequencies between any pair of gyrotrons are not controllable .

The LHD, which is the large helical device for fusion plasma experiments in NIFS, has ICH and EC systems for high rf power and long pulse operations with HIBP diagnostic system using 6 MeV Au^+ ion beam [19]. We would like to expect the excellent study of wave physics using such a next generation HIBP system in LHD.

CONCLUSION

The new diagnostic method is proposed for direct measurements of electromagnetic field pattern of fast wave in toroidal plasmas. The two frequency fast waves excites a fluctuating ponderomotive force at the beat wave frequency. The electrons and ions respond to the ponderomotive force and produce a potential fluctuation. The potential fluctuation pattern at the beat wave frequency has been measured with HIBP system during the fast wave pulses in JFT-2M FWCD experiments. The measured levels of potential fluctuations decrease with increasing an electron temperature, consistent with the improvement of wave absorption. The difference between the experimental results (flat profile) and the theoretical predictions (standing wave structure) can be explained by the spatial resolution of HIBP system. The measured potential fluctuation levels are similar to the theoretical predictions. The feasibility of this diagnostic method in large fusion devices depends on the wave absorption and the sensitivity of HIBP.

ACKNOWLEDGMENTS

The authors would like to acknowledge Drs. T. Ogawa, T. Ido, and H. Kawashima for their contributions on experiments. We would like to acknowledge Dr. A. Fukuyama for permission to use the full wave code (TASK/WM) developed by himself. We would like to thank K. Kikuchi and the JFT-2M team for operating the FWCD system and the tokamak. This work was partly supported by the Grant-in-Aid program of the Ministry of Education Science, Sports and Culture of Japan.

REFERENCES

[1]   Y. Uesugi, T. Yamamoto, H. kawashima, et al., *Nucl. Fusion* 30 (1990) 297.

[2]   M. Ono, *Phys. Plasmas* 2, (1995) 4075.

[3]   Fukuyama, 18th IAEA Nuclear Fusion Conference, "Global Analysis of ICRF Waves and Alfvén Eigenmodes in Toroidal Helical Plasmas", *THP2/26* (2000).

[4]   Forest, et al., *Phys. Rev. Lett.* 73 (1994) 2444.

[5]   Mazurenko, et al., "Fluctuation and fast wave measurements by the phase constrast imaging on Alcator C-Mod", 42nd APS DPP Meeting, Quebec, UP1.105 (2000).

[6]   M. Saigusa, S. Kanazawa, T. Ogawa, H. Kawashima, K. Kikuchi, T. Ido, A. Fukuyama, *Nucl. Fusion*, 42(2002)412-417.

[7]   M. Ono, K. L. Wong, G. A. Wurden, *Phys. Fluids* 26, 298-309(1983).

[8]   J. R. Wilson, R. E. Bell, S. Bernabei, K. Hill, J. C. Hosea, et al., *Physics of Plasmas* 5, 1721 (1998).

[9]   A.K. Ram and S.D. Schultz, *Physics of Plasmas* 7, 4084 (2000).

[10] ITER Director, *Summary of the ITER Final Design Report*, July 2001, 21 (2001).

[11] F. F. Chen, *Introduction to Plasma Physics*, Plenum Press, New York (1974).

[12] Guy Dimonte, B.M. Lamb, and G.J. Morales, *Phys. Rev. Lett.* 48 (1982) 1352.

[13] H. Motz and C.J.H. Watson, "The radio-frequency confinement and acceleration of plasmas", *Advances in Electronics and Electron Physics*, Academic Press (1967) 153.

[14] J. C. Hosea, F.C. Jobes, R.L. Hickok, and A.N. Dillis, *Phys. Rev. Lett.* 30 (1973) 839.

[15] Fukuyama, et al., *Comp. Phys. Rep.* 3&4 (1986) 137.

[16] MOELLER, C.P., CHIU, S.C., PHELPS, D.A., in *Proc. Europhysics Top. Conf. on Radiofrequency Heating and Current Drive of Fusion Devices*, Brussels, 1992 (European Physical Society, 1992), Vol. 16E, (1992) 53.

[17] T. Ogawa, T. Ogawa, K. Hoshino, S. Kanazawa, M. Saigusa, T. Ido, H. Kawashima, N. Kasuya, T. Takase, H. Kimura, Y. Miura, K. Takahashi, C.P. Moeller, R. I. Pinsker, C.C. Petty and JFT-2M group,et al., *Nucl. Fusion* 41, (2001)1767.

[18] T. Ido, Y. Hamada, A. Nishizawa, Y. Kawasumi, Y. Miura, and K. Kamiya, *Rev. Sci. Instrum.*, 70 (1998) 955.

[19] T. Ido, A. Nishizawa, Y. Kawasumi, K. Tsukada, S. Kato, M. Yokota, H. Ogawa, T. Inoue, " Development of a Heavy Ion Beam Probe for LHD", *Annual Report of National Institute for Fusion Science*, April 2002-March 2003, (2003)165.

INDEX

global scale, 9
glow discharge, ix, 111, 112, 114, 115, 117, 118, 119, 120, 121, 123, 124, 125
grain size, 99
graphite, ix, 111, 113, 115, 117, 119, 120, 121, 124, 146
grounding, 71, 79, 84
growth, 104, 146
growth rate, 104

H

harmful effects, 68
heat capacity, 144
heating rate, 115
height, 159
helium, x, 73, 111, 116, 117, 118, 119, 120, 121, 123, 124, 125
helium ions integrated, vii, 1, 3
hemisphere, 10
history, 12, 87, 88, 142
hybrid, ix, 67, 70, 97, 168, 175
hydrogen, xi, 69, 70, 86, 88, 89, 90, 91, 95, 112, 116, 118, 121, 123, 124, 125, 141, 144, 148, 149, 150, 151, 152, 159, 160, 161, 162, 164, 165
hydrogen atoms, 88, 161
hydrogen gas, 150, 160

I

ice pellets, x, 141, 142
image, 3, 4, 11, 12
IMAGE spacecraft, vii, 1
images, vii, 1, 8, 10, 11
impurities, x, 70, 83, 95, 111, 112, 113, 115, 117, 118, 121, 123, 125, 147, 148, 149, 164
inertia, 80, 170
injections, 154
insulation, 112
integration, 98
interface, 99
interference, 74, 147
International Thermonuclear Experimental Reactor (ITER), xi, 142
interplanetary magnetic field, vii, 1, 7
inversion, 147, 152
ion implantation, 118
ionization, 80, 86, 91, 92, 95, 168, 169
ionosphere, vii, 1, 10, 11, 13

ions, vii, viii, 1, 3, 10, 11, 69, 72, 75, 78, 80, 83, 84, 85, 86, 95, 96, 99, 105, 106, 117, 121, 125, 168, 169, 171, 172, 178, 181, 182, 183
iron, ix, 111, 124
Islam, v, 106, 107, 108, 109
isotope, 148
issues, 68

J

Japan, 67, 106, 108, 111, 127, 167, 181, 183

K

kinetics, 146

L

large helical device, LHD, ix, 111, 112
lead, ix, 69, 125, 148, 164
lens, 160
Levy distributions, viii
light, 74, 112, 149, 160, 162, 169
linear systems, 68
lithium, 148
luminosity, 3, 4
lying, 99

M

magnetic field, vii, x, xi, 1, 7, 69, 71, 72, 73, 75, 76, 78, 80, 81, 83, 86, 88, 93, 94, 95, 97, 99, 102, 105, 106, 113, 119, 141, 144, 154, 165, 167, 168, 169, 170, 171, 172, 173, 174
magnetic field effect, 75, 83, 93
magnetic filed configuration (MFC), ix, 67
magnetic fusion, 68
magnetic moment, 86
magnetosphere, 2, 10, 11, 13, 14
magnitude, 11, 80, 101, 124, 142, 159, 164
majority, 181, 182
mapping, 73
Maryland, 127
mass, x, 14, 96, 99, 112, 115, 120, 125, 141, 142, 145, 149, 171, 172
mass loss, 14
Master Equation, ix
material probe study, ix, 111
materials, 70, 73

N

O

P

Q

R